SAN___ON W___R

ARCO CIVIL SERVICE BOOKS

TEST PREPARATION

Accountant / Auditor
ACWA: Administrative Careers With America
Air Traffic Controller Qualifying Test
Air Traffic Controller Training Program
American Foreign Service Officer
Assistant Accountant
Auto Mechanic / Automotive Serviceman
Beginning Clerical Worker
Bookkeeper / Account Clerk
Border Patrol Agent
Building Custodian / Building Superintendent /
    Custodian Engineer
Bus Operator / Conductor
Case Worker
CIA Entrance Examination
Computer Specialist GS 5-9
The Corey Guide to Postal Exams
Correction Officer
Correction Officer Promotion Tests
Court Officer / Senior Court Officer / Court Clerk
Distribution Clerk, Machine
Drug Enforcement Agent
Electrician / Electrician's Helper
Emergency Dispatcher / 911 Operator
FBI Entrance Examination
Federal Clerk / Steno / Typist
File Clerk / General Clerk
Fire Department Lieutenant / Captain /
    Battalion Chief
Firefighter
Gardener / Grounds Maintenance Worker
Investigator / Claim Examiner
Librarian
Machinist / Machinist's Helper
Mail Handler / Mail Processor
Maintenance Mechanic
Maintenance Worker / Mechanical Maintainer
Mark-up Clerk / Clerk Typist /
    Clerk Stenographer—U.S. Postal Service
Office Aide
Office Associate
Plumber / Steam Fitter
Police Administrative Aide
Police Officer
Police Sergeant / Lieutenant / Captain
Postal Exams Handbook

Post Office Clerk / Carrier
Preparacion Para El Examen de Cartero
Principal Administrative Associate /
    Administrative Assistant
Principal Clerk / Principal Stenographer
Probation Officer / Parole Officer
Railroad Clerk
Rural Carrier
Sanitation Worker
School Secretary
Senior Clerical Series
Special Agent
Special Officer / Senior Special Officer /
    Bridge and Tunnel Officer
Staff Analyst / Associate Staff Analyst
State Trooper / Highway Patrol Officer /
    State Traffic Officer
Track Worker
Traffic Enforcement Agent
Train Operator / Tower Operator / Assistant
    Train Dispatcher
Transit Electrical Helper

CAREERS / STUDY GUIDES

Civil Service Administrative Tests
Civil Service Arithmetic and Vocabulary
Civil Service Clerical Promotion Tests
Civil Service Handbook
Civil Service Psychological and Psychiatric
    Tests
Civil Service Reading Comprehension Tests
Civil Service Tests for Basic Skills Jobs
Complete Guide to U.S. Civil Service Jobs
Federal Jobs for College Graduates
General Test Practice for 101 U.S. Jobs
Homestudy Course for Civil Service Jobs
How to Get a Clerical Job in Government
How to Pass Civil Service Oral Examinations
New York City Civil Service Job Guide
101 Challenging Government Jobs for
    College Graduates
Practice for Clerical, Typing and
    Stenographic Tests
SF 171: The Federal Employment
    Application Form
Supervision
You as a Law Enforcement Officer

AVAILABLE AT BOOKSTORES EVERYWHERE

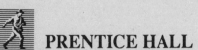

PRENTICE HALL

SIMON & SCHUSTER / A PARAMOUNT COMMUNICATIONS COMPANY

SANITATION WORKER

Hy Hammer

Prentice Hall
New York • London • Toronto • Sydney • Tokyo • Singapore

Seventh Edition

Copyright © 1992, 1983 by Arco Publishing,
 a division of Simon & Schuster, Inc.

 PRENTICE HALL

15 Columbus Circle
New York, NY 10023

An Arco Book

Prentice Hall and colophons are
registered trademarks of Simon & Schuster, Inc.

Manufactured in the United States of America

1   2   3   4   5   6   7   8   9   10

Library of Congress Cataloging-in-Publication Data

Hammer, Hy.
  Sanitation worker / Hy Hammer.
     p. cm.
  ISBN 0-13-788316-1
  1. Refuse and refuse disposal--Examinations, questions, etc.
2. Street cleaning--Examinations, questions, etc. 3. Snow removal-
-Examinations, questions, etc. I. Title.
TD791.H25 1992   91-44662
628.4'076--dc20   CIP

CONTENTS

INTRODUCTION

Sanitation workers perform one of the least glamorous but most important public services. If you have ever lived through a sanitation workers' strike, you know how quickly you learned to appreciate sanitation workers. Of course, garbage collection is only one of many services performed by sanitation workers.

Garbage and other trash collection is perhaps the first job that comes to mind, but sanitation workers also empty collection baskets on street corners and keep the streets clean. In congested areas, sanitation workers may sweep the gutters with big brooms, dustpans, and large collection barrels. In residential areas, they drive sweepers and flushers to keep the streets clean. In regions of measurable snowfall, they salt, sand, and plow.

Sanitation workers not only collect; they also dispose of the material they collect. Sanitation workers transport refuse to dumps, incinerators, landfills, treatment plants, and garbage barges. They transport collected snow to open areas for dumping or dump the snow directly into rivers. In some localities, sanitation workers operate garbage barges or run the treatment plants. With the drive to cut down on air and water pollution, sanitation workers are very much involved in ecology and ''saving the environment.'' As new methods are invented for dewatering refuse and turning garbage into safe, usable fertilizer, sanitation workers are on the front lines of progress.

Even at the entry level, sanitation workers are well paid. Since the work is not glamorous and may, at times, be truly unpleasant, the pay scale for sanitation workers is higher than that of others entering the unskilled work force. At most times, the work is regularly scheduled. Most sanitation workers start work very early in the day. This may mean getting up earlier than you like to but, remember, the worker who starts early, ends early. This means that most sanitation workers have a good part of the afternoon to themselves. They can go to afternoon ballgames or cultivate other interests. If not scheduled for weekend work, they can get an early start for weekend holidays and can beat the traffic.

When there are emergencies—a parade that creates extra litter for street cleaners or a heavy snowstorm—sanitation workers are called upon to put in overtime and are paid at overtime rates. Overtime is a nice ''extra'' that cannot be counted into the budget because you never know when and how much. It can be a lifesaver for clearing long-term debts or you can save it for vacations.

Beyond the advantages of performing a public service, good pay, and agreeable hours, sanitation work can be fun. Sanitation crews that work together tend to develop close friendships and enjoy working together. While there are rules to be followed, there is no minute-to-minute supervision. Sanitation workers have a higher degree of freedom than people who work in factories or offices under the steady eye of a boss.

Advancement for a sanitation worker probably begins with choosing the hours and route that suit your needs rather than being assigned shifts when you are the newest worker in the department. With experience and time on the job, the worker can become chief of a crew and then move into positions of greater responsibility in the department. There is an upward path for sanitation workers who want to remain with the department for their careers.

In a recent newspaper interview, the commissioner of a sanitation district in a large city spoke fondly of his many years with the sanitation department. He enjoyed being the leader on a truck saying, ''That's probably the best part of the job.'' He has been with the department 39 years and, as a commissioner, has a responsible but not as physically demanding a job. He can't think of any reason to retire and he highly recommends the life of a sanitation worker.

INTRODUCTION

PART ONE

APPLYING AND STUDYING

BECOMING A SANITATION WORKER

Sanitation work is popular. Sanitation workers are always needed, so the work offers real job security along with good pay. The combination of job security, good pay, and no educational requirements draws many applicants for every opening. In 1990, 102,000 people applied for a single examination in hope of becoming sanitation workers in New York City. Obviously, if so many people want to work for the sanitation department, there must be some selection process. There must be a way to choose who is to be hired.

Most sanitation departments require applicants to pass both a written test and a physical performance test. Under federal law, all sanitation drivers must be able to pass a commercial driver's licensing test. Some departments are organized so that workers have specialized duties and not everyone drives. In these departments the commercial driver's license is not part of the selection process. Most sanitation departments, however, want all employees to be able to do every job. These departments require that job applicants be eligible for and able to pass a commercial driver's licensing test. No department requires that you have the commercial license at application time, only that you be qualified to earn it soon.

The written test for sanitation worker applicants does not ask any questions about department rules or about how to do the work. No one expects you to know the job before you have been trained or before you have actually done the work. What the exam looks for is some common sense, an ability to understand what you read, and the ability to follow directions. Many sanitation worker exams will also be looking for the ability to read simple tables and do some basic arithmetic. The written exam is not difficult, but it can be frightening if you have not taken an exam recently. Do not worry about it. This book includes five model examinations for practice. These are not actual Sanitation Department exams. The real exams are not made public. But these exams are very close to the real thing. We have spoken with people who have taken recent sanitation worker exams and have constructed exams on the basis of their reports. The five practice exams are not all alike because not every city or town gives the same exam. Overall, though, these practice exams should give you a good idea of the types of questions you will have to answer, the level of difficulty, and the time limits. If you can handle these practice exams, you can pass your sanitation worker exam.

The physical performance test does not test for any kind of knowledge; it tests for physical strength, speed, and endurance. This book describes three different physical performance tests. It then gives you a full chapter of instruction on how to build yourself up for the test. Start preparing early so that you will be ready when you are called to the exam.

The commercial driver's license requirement is brand new. Under federal law, all drivers of sanitation collection trucks (as well as other heavy vehicles) must hold the commercial license. Each state is in charge of testing for and granting the commercial license. Each state can set its own schedule for phasing in the license; however, by 1993 no one will be permitted to drive a commercial vehicle without the proper license.

The commercial licensing law is so new that most sanitation departments have not yet set up procedures for dealing with it. Most departments will require that you have a regular driver's license before appointment, and many will require that you have a commercial learner's permit at that time. Then they will allow a period of time in which you can learn to drive the bulky, heavy equipment before you must earn the commercial license. Your state department of motor vehicles can give you a booklet describing the rules for obtaining the license in your state. The booklet will be filled with information that you must know to pass the test. Ask wherever you apply for a Sanitation Department position—the personnel office of your city or town or the Sanitation Department itself—how the commercial license law applies to you and how you should prepare yourself.

USING THIS BOOK

You have made a good start toward getting a job with the Sanitation Department by buying this book. But, buying the book is not enough; you must now use it.

Now that you have read the introduction, you know what the job of the sanitation worker is all about and you are eager to try out for it. You have read the chapter on becoming a sanitation worker and have found that it is not all that easy. Not everyone who wants to be a sanitation worker can be one.

Continue reading Part One and find out about application procedures; how to read an announcement; and how to follow through with the application. Then turn your attention to the section on test-taking techniques. Read this section before you try the model exams in the book, and then reread it on the day before you go for your written exam. You can't take the book with you to the exam, but you can take the helpful hints along in your head if you go over them the day before.

Next move on to Part Two where you will find five full-length model exams. These exams have been specially constructed to give you practice with many different kinds of sanitation worker exam questions. Not all kinds of questions appear on every exam. Since you do not know what kind of questions will be on your exam, work your way through every question on all five exams. Take each exam in one sitting; do not break any exam into halves or smaller portions. Stick to time limits. Do not peek at the answers while taking an exam. Answer as best you can and learn from your mistakes. Each exam is followed by an answer key and explanations of the correct answers. Read the explanations to find out why the right answer is right. Read the explanation even if you got the answer right. The explanation may make the point even clearer.

The exams are all very different, so you cannot compare your score on any one exam with your score on any other. If you feel that your score was too low on any of the exams, repeat those exams after you have completed the book. There are extra answer sheets at the back of the book that you can use for repeating exams.

In Part Three of this book you will learn about medical requirements and physical performance tests. We offer you three different physical performance tests given in different places at different times. We cannot predict your test, but these three will alert you to what you must be prepared to do. The remainder of Part Three is devoted to helping you prepare your body for the physical performance test. Skim through this chapter; then go back and follow through on the fitness program one step at a time. Start now so that you will be in perfect shape by the time you take a physical performance test. Building yourself up is a slow process. Do not wait until the exam is announced to begin.

Part Four ends the book with important information that you must have to take your exams and to perform your duties as a sanitation worker.

APPLYING FOR THE JOB

If you want to join the Department of Sanitation, call the personnel office of your city or town and ask how to go about it. Very few places have open applications. Usually, filing applications is limited to a set period and examinations are given at one time to all applicants. In some places this may happen every year or even twice a year; in other places the sanitation worker examination is offered only every four years or so. You must ask when, where, and how to file. If the examination will not be open fairly soon, you may decide to look for another job and keep your eyes and ears open for an announcement of the opening of the filing period.

When a filing period is about to open, an announcement is made on personnel and sanitation department bulletin boards and on bulletin boards in state employment offices and job banks and in public libraries and municipal buildings. Civil service newspapers and many local newspapers will publish notice of the open filing period and local radio stations may broadcast this information as a public service.

The announcement includes information that is very important to applicants. It tells the limits of the dates of the filing period and where to file. It tells what form must be used for the application and how to pay the filing fee, if any. The announcement describes the duties of the job itself and mentions special working conditions.

The announcement also describes eligible applicants in terms of education, medical and physical requirements, residency requirements, and special licenses needed. If there are age limits, the announcement will specify these as well. Under the new federal requirements, all drivers of heavy sanitation department vehicles must hold a Class B commercial driver's license. You must be 21 years of age to be issued a Class B commercial driver's license, though a younger licensed driver can get a learner's permit. If your city or town will appoint only candidates who are qualified to do every job, then the effective age for appointment must be 20 and a half. However, the selection process takes some time. The minimum age for application may be quite a bit lower and some locations may appoint younger workers to perform non-driving duties for the first few years. Read your announcement carefully for minimum age of applicants and minimum age for appointment. If the announcement is not clear, call the Sanitation Department or personnel office.

No two locations put out exactly the same announcement; each is different in detail but all contain similar information. The New York City Sanitation Worker announcement that follows is a good example of a job announcement. The information under each heading is specific to New York City, but the headings themselves show you what kind of information an announcement gives.

APPLICATION FEE: $
ANNUAL SALARY: $

Job Announcement

Sanitation Worker Exam. No. 9147

APPLICATION PERIOD: From April 30, 19– through May 19, 19–. Application forms will be available at the Department of Personnel and at Department of Sanitation garages and District Offices. Application forms with money order payment of application fee must be submitted by *MAIL ONLY* to the Applications Section. On the lower left-hand corner of the envelope, write "Sanitation Worker." Properly completed Application Forms must be *received* by the last date of the application period. Date of receipt, rather than date of postmark, will be controlling.

MEDICAL STANDARDS: Eligibles will be rejected for any medical condition which impairs their ability to perform the duties of this job or which may reasonably be expected to render them unfit to continue to perform those duties. A drug screening test will be conducted for this position. Medical Standards are available at the Application Section of the Department of Personnel.

MINIMUM AGE REQUIREMENT: Eligibles must have reached the age of twenty and one-half by the date of appointment, unless they hold a class 1, 2, or 3 license.

LICENSE REQUIREMENTS: At the time of appointment, eligibles must have at least a valid Class D New York State driver license and a Class B learner's permit. Serious moving violation(s), driver license suspension, accident record, or other indication of poor driving ability may disqualify eligibles. Appointees will be required to obtain a New York State Class B driver license after training.

JOB DESCRIPTION: Under direct supervision, performs the work and operates the equipment involved in street cleaning, waste collection, recycling collection, snow removal, and waste disposal; performs related work.

EXAMPLES OF TYPICAL TASKS: Follows route sheet of assigned work area; loads and unloads waste materials and bulk items; sweeps and cleans city streets; participates in special cleaning and collection work details; inspects and operates department vehicles and equipment; during winter months, attaches and operates snow plows, removes snow and ice from city streets and arterial roadways, and spreads salt and sand; monitors collected waste for hazardous or toxic materials; secures and covers barges; uses operational manuals and instructional materials; prepares necessary forms and reports; makes log entries; moors and shifts barges.

Eligibles appointed to this position may be assigned to work rotating shifts including nights, Saturdays, Sundays and holidays.

RESIDENCY REQUIREMENT: The New York State Public Officers Law requires that any person employed as a Sanitation Worker in the New York City Department of Sanitation must be a resident of the City of New York or of Nassau or Westchester Counties.

ENGLISH REQUIREMENT: All candidates must be able to understand and be understood in English. A qualifying English oral will be given by the Department of Personnel to all candidates who, in the opinion of the appointing officer, do not meet this requirement.

PROOF OF IDENTITY: Under the Immigration Reform and Control Act of 1986, you must be able to prove your identity and your right to obtain employment in the United States prior to employment with the City of New York.

PROBATIONARY PERIOD: The probationary period is one year. As part of the probationary period, probationers must successfully complete a prescribed training course.

INVESTIGATION: At the time of investigation and at the time of appointment, candidates must present originals or certified copies of all required documents and proof, including but not limited to proof of date of birth by transcript of record of the Bureau of Vital Statistics or other satisfactory evidence and proof of any military service. Any willful misstatement or failure to present any required documents will be cause for disqualification.

SPECIAL ARRANGEMENTS: Accommodations are available for certified disabled applicants. Applications for accommodations must be submitted as early as possible and in no event later than 30 work days before the test or part of a test for which accommodation is requested.

A candidate who is prevented from participating in the examination on the test date because of religious beliefs may request a special examination in accordance with the provisions of Personnel Director Rule 4.4.6.

TEST DESCRIPTION: Physical test, competitive, weight 100; written test, qualifying.

Physical Test—Your score will be determined by the physical test. A score of at least 70% on the physical test is needed to pass. A description of the physical test will be provided at a later date.

Written Test—All applicants will be called to the qualifying written multiple-choice test. The pass mark will be 70%.

QUALIFYING PHYSICAL AND MEDICAL: Eligibles must meet the medical and physical fitness standards to be established and posted on the Department of Personnel Bulletin Board. Periodic reexaminations of employees in this title may be required to ensure they are in good health.

PHYSICAL TEST: Medical evidence to allow participation in the physical test may be required and the Department of Personnel reserves the right to exclude from the physical test eligibles who are medically unfit.

TEST DATE: The written test is expected to be held on September 22, 19–.

There is no universal application form. Your application form may be as simple as the New York City application on pages 12–13. This form asks only for name, address, social security number and certain questions about veteran's preference status and special needs. It also asks for exam number and exact job title, both of which can be found on the announcement.

Other application forms may require more information. They may ask about age, birthdate, education, driver's license, citizenship or permanent residence status, previous work experience, and if you have ever worked for the government. Whatever the form asks, answer truthfully and completely. The application form is the basis on which the hiring agency will decide whether or not to give you the test. In New York City, everyone who applies can take the test. New York City checks on job eligibility after the examination, not before. If your application form is more detailed, your locality may check the form first and rule applicants in or out on the basis of eligibility factors like age, driver's license, or place of residence.

APPLICATION FOR OPEN COMPETITIVE EXAMINATION
(See Instructions on Reverse Side)

CITY OF NEW YORK
DEPARTMENT OF PERSONNEL
APPLICATION SECTION
49 THOMAS STREET
NEW YORK, NEW YORK 10013

EXAM NO.

YOUR SOCIAL SECURITY NO. (check your card before entering your number)

EXAM TITLE

FEE (See Item 3 on reverse)

$

YOUR FIRST NAME

Middle Initial

YOUR LAST NAME

YOUR STREET ADDRESS (INCLUDE APT. NO., BLDG. NO., OR CARE OF)

CITY OR TOWN

STATE

ZIP CODE

IF YOU LIVE IN N.Y.C., CHECK THE BOROUGH THAT YOU LIVE IN.

MANHATTAN M   BRONX X   BROOKLYN K   QUEENS Q   STATEN ISLAND R

CHECK ALL BOXES THAT APPLY TO YOU.

V I CLAIM VETERANS' CREDIT. (SEE ITEM 7 ON REVERSE.)

D I CLAIM DISABLED VETERANS' CREDIT. (MINIMUM 10% DISABILITY) (SEE ITEM 7 ON REVERSE.)

S I AM A SABBATH OBSERVER. (SEE ITEM 4 ON REVERSE.)

H I HAVE A HANDICAP REQUIRING SPECIAL TESTING ACCOMMODATIONS. (SEE ITEM 5 ON REVERSE.)

DECLARATION: I declare that the statements on this form are true. I further declare that if I have made a claim for Veterans' or Disabled Veterans' credit that I meet the requirements for such credits as described on the reverse side. I am aware that I must prove that I am entitled to the credit I claim or my employment may be terminated.

Your Signature _____   Date _____

REMINDER

- DID YOU ENCLOSE THE CORRECT FEE?
- DID YOU FILL OUT COMPLETELY AND ATTACH ALL OTHER FORMS REQUIRED BY NOTICE OF EXAMINATION?
- IS YOUR SOCIAL SECURITY NUMBER CORRECT?

- DID YOU SIGN YOUR NAME?
- DID YOU ANSWER EVERY QUESTION?
- DID YOU GIVE YOUR COMPLETE ADDRESS?
- DID YOU INCLUDE YOUR CORRECT ZIP CODE?

AN INCOMPLETE APPLICATION MAY NOT BE PROCESSED. BE SURE TO INDICATE YOUR FULL NAME, SOCIAL SECURITY NUMBER, ADDRESS AND ZIP CODE.

BE SURE TO READ THE INSTRUCTIONS ON THE REVERSE SIDE.

DP-848C (R. 3/89)

INSTRUCTIONS FOR FILLING OUT THIS APPLICATION FORM

NOTE: You should only apply for an examination if you meet the qualification requirements set forth in the Notice of Examination.

Be sure to read the Notice of Examination carefully before completing this form. Fill in all requested information clearly, accurately and completely.

1. **FORMS**
If the Notice of Examination calls for Experience Form A and/or other forms, these forms must be included with your application, or it will not be accepted.

2. **ADDRESS**
Give your full mailing address, including building, apartment number, or "in care of" information where needed. If you change your address after applying, write to the Examining Service Division of the Department of Personnel, Room 216, 220 Church Street, New York, N.Y. 10013, with your new address, plus your social security number and the number and title of the examination for which you applied.

3. **FEE**
Enclose the correct fee. The amount of the fee is stated in the Notice of Examination. CHECKS WILL NOT BE ACCEPTED. If paying by mail, use money order only made out to the N.Y.C. Department of Personnel. DO NOT MAIL CASH. Write your name, social security number, and the examination number on the front of the money order. A fee is not charged if you are a N.Y.C. resident receiving public assistance from the N.Y.C. Department of Social Services. To have the fee waived you must present or enclose a clear photocopy of your current Medicaid Card. The photocopy must accompany the application, even if it is filed in person.

4. **SABBATH OBSERVER**
If, because of religious belief, you cannot take the test as scheduled, you must request an alternate date by coming or writing to: Examining Service Division, Room 216, N.Y.C. Department of Personnel, 220 Church Street, New York, N.Y. 10013 no later than 15 workdays prior to the test date. All requests must indicate the examination title, examination number, and your social security number, and must be accompanied by a signed statement on letterhead from your religious leader certifying that your religious observance prohibits you from taking the test as scheduled.

5. **DISABILITY**
If you have a disability which will interfere with your ability to take this test without special accommodations, amanuensis, or other assistance, you must submit a written request for specific special accommodations to the Examining Service Division, Room 216, N.Y.C. Department of Personnel, 220 Church Street, New York, N.Y. 10013. The request must be received no later than 30 workdays prior to the test date. A physician or agency authorized for this purpose must corroborate the specific nature of your disability and must justify the need for the special accommodations you request. For more details, consult Regulation E.10 of the General Examination Regulations, available from the Application Section of the Department of Personnel.

6. **MAIL AND FURTHER INFORMATION**
Check the Notice of Examination to see if applications are accepted by mail. If you mail your application, use a 4¼"x9½" (legal size) envelope, and address it to:

N.Y.C. Department of Personnel
Application Section
49 Thomas Street
New York, New York 10013

YOUR APPLICATION MUST BE RECEIVED BY THE N.Y.C. DEPARTMENT OF PERSONNEL BY THE LAST DAY OF FILING. For information, see the General Examination Regulations, available from the Application Section of the Department of Personnel.

7. **VETERANS' CREDIT**
FOR VETERANS' OR DISABLED VETERANS' CREDIT YOU MUST:

a. Have served on active duty, other than for training purposes, in the Armed Forces of the United States during:
- December 7, 1941 - September 2, 1945; or
- June 26, 1950 - January 31, 1955; or
- January 1, 1963 - May 7, 1975; and
b. Be a resident of New York State at the time of list establishment; and
c. Have an honorable discharge or have been released under honorable conditions.
d. For Disabled Veterans' Credit, in addition to a, b, and c above, at the time the list is established, you must receive or be entitled to receive at least 10% compensation from the V.A. for a disability incurred in time of war. The V.A. must also certify that the disability is permanent. If it is not permanent, you must have been examined by the V.A. within one year of the establishment of the list.

NOTES
1. You may use Veterans' or Disabled Veterans' Credit only once after January 1, 1951 for appointment or promotion from a City, State or County civil service list.
2. The above is only a summary of necessary conditions. The complete provisions are contained in statutory and/or decisional law.

TEST-TAKING TECHNIQUES

Many factors enter into a test score. The most important factor should be ability to answer the questions correctly. Ability to answer exam questions should be closely related to ability to learn and to perform the duties of the job. Assuming that you have this ability, it is important to know what to expect on the exam and be familiar with techniques of effective test-taking. This will ease the anxiety which might interfere with concentration and should increase speed and efficiency, enabling you to answer more questions and thereby raise your score.

On Examination Day

On the examination day assigned to you, allow the test itself to be the main attraction of the day. Do not squeeze it in between other activities. Arrive rested, relaxed, and on time. In fact, plan to arrive a little bit early. Leave plenty of time for traffic tie-ups or other complications which might upset you and interfere with your test performance. Remember to bring a few sharpened #2 pencils with clean erasers, positive identification with your picture or at least your signature, and whatever admission ticket or other papers you were instructed to bring. You will not be admitted to the exam if you have forgotten the required documents.

In the test room the examiner will hand out forms for you to fill out. He or she will give you instructions that you must follow in taking the examination. The examiner may distribute pencils for marking the answer sheet and will tell you how to fill in the grids on the forms. Time limits and timing signals will be explained. If you do not understand any of the examiner's instructions, ask questions. Make sure that you know exactly what to do.

At the examination, you must follow instructions exactly. Fill in the grids on the forms carefully and accurately. Filling in the wrong grid may lead to loss of veterans' credits to which you may be entitled or to an incorrect address for your test results. Do not begin until you are told to begin. Stop as soon as the examiner tells you to stop. Do not turn any pages until you are told to. If your exam is in parts, do not go back to any parts you have already completed. Any infraction of the rules is considered cheating. If you cheat, your test paper will not be scored, and you will not be eligible for appointment.

Using The Answer Sheet

Your exam will probably be machine scored. You cannot give any explanations to the machine, so you must fill out the answer sheet clearly and correctly.

1. Blacken your answer space completely. ● is the only correct way to mark the answer sheet. ◖, ⊗, ⊘, and ∅ are all unacceptable. The machine might not read them at all.
2. Mark only the one answer for each question. If you mark more than one answer you will be considered wrong even if one of the answers is correct.

3. If you change your mind, you must erase your mark. Attempting to cross out an incorrect answer like this ✖ will not work. You must erase any incorrect answer completely. An incomplete erasure might be read as a second answer.

4. All of your answers should be in the form of blackened spaces. The machine cannot read English. Do not write any notes in the margins.

5. MOST IMPORTANT: Answer each question in the right place. Question 1 must be answered in space 1; question 23 in space 23. If you should skip an answer space and mark a series of answers in the wrong places, you must erase all those answers and do the questions over, marking your answers in the proper places. You cannot afford to use the limited time in this way. Therefore, as you answer *each* question, look at its number and check that you are marking your answer in the space with the same number.

What About Guessing?

You may be wondering whether or not it is wise to guess when you are not sure of an answer. Scoring of almost all civil service exams is on the basis of "rights only." The more questions you answer correctly, the higher your score. On these exams there is no penalty for a wrong answer. If you guess incorrectly, you have nothing to lose; but if you do not mark an answer, you cannot get a point. Ask to be sure about your exam. If there is no penalty for guessing wrong, by all means guess.

Obviously, an educated guess is worth more than a wild guess. When you are not certain of an answer, read the question and the answer choices very carefully. Eliminate those choices which are certainly wrong. Then try to reason from the remaining choices. Narrow the field as much as you can, then guess from among those choices which are left. The odds of guessing right improve as the number of choices from which you guess gets smaller.

Since you cannot score a point without marking an answer and since a wrong answer does not count against you, it is foolish to leave any blank spaces. Keep track of time. When you notice that time is about to run out, use the last few seconds to complete the exam part, even without reading the questions. Just choose a letter and mark all the remaining blanks with the same answer. According to the law of averages, you should get some portion of those answers right. If yours is the unusual exam which deducts points for wrong answers, do not fill in leftover spaces at time-up.

How Is The Exam Scored?

When the exam is over, the examiner will collect test booklets and answer sheets. The answer sheets will be sent to a central test center where a machine will scan your answers and mark them right or wrong. Then your raw score will be calculated. Your raw score is the number of questions you answered correctly.

This is not your final score. Raw scores are converted by formula to scaled scores, usually on a scale of 1 to 100.

In some locations, the written exam is competitive and candidates with the highest scores have the best chance of being hired. In other places, the written exam is qualifying. This means that the applicant qualifies for further consideration just by passing the written exam. Where the written exam is qualifying, appointment is usually based on a competitive physical performance exam. Other localities combine scores on written and performance exams into a single competitive score.

Except where appointment is made by lottery from among all passing applicants, appointment is made first from among the top scorers. If you are entitled to veterans' service points, these are added to the score you earned. Veterans' points are added only to passing scores. A failing score cannot be brought to passing level by veterans' points, but if you pass, veterans' points can help you get the job.

Recap of the Rules

1. Read every word of the instructions. Read every word of every question. Little words like *not, all* or *except* may make a big difference in determining the correct answer.
2. Read all the answer choices before you mark your answer. It is statistically true that most errors are made when the correct answer is the last choice. Too many people mark the first answer that seems correct without reading through all the choices to find out which answer is best.
3. Mark your answers by completely blackening the answer space of your choice.
4. Mark only **one** answer for each question, even if you think that more than one answer is correct. You must choose only one. If you mark more than one answer, the scoring machine will consider you wrong.
5. If you change your mind, erase completely. Leave no doubt as to which answer you mean.
6. If you figure math in the margins of the test booklet, don't forget to mark the answer on the answer sheet. Only the answer sheet is scored. A correct answer in the margin of the test booklet does not count.
7. Check often to be sure that the question number matches the answer space, this will ensure that you have not skipped a space by mistake.
8. Guess as intelligently as you can, but be sure to answer all the questions.
9. Stay alert. Be careful not to mark a wrong answer from lack of concentration.
10. Do not panic. The person beside you who is way ahead of you may be making lots of mistakes. Just do your best.
11. Check and recheck if time permits. If you finish early, check to be sure that each question is answered in the right space and that there is only one answer for each question. Return to the difficult questions and rethink them.

Good Luck!

PART TWO

THE WRITTEN EXAMS

SANITATION WORKER EXAM INSTRUCTIONS

These are the official instructions to accompany the announcement on page 9. There is always some variation from exam to exam, and each department gives its own exam with its own instructions, but these give you a good idea of what you will be told before you begin. The model exam that begins on page 23 is based on the exam that went with these instructions.

SANITATION WORKER EXAM NO. 9147

QUALIFYING WRITTEN TEST:
70% REQUIRED

Time Allowed: 2½ Hours

FIRST SIGNAL: Test Question Booklet is given out. *Read instructions on this page, Page 2 and Page 3.* Fill in the information asked for at the top of this page.

SECOND SIGNAL: Turn to Page 4 and begin work. Check to make sure that this test booklet has *60* questions. Time allowed includes time for announcements and recording your answers on your OFFICIAL ANSWER SHEET and on your Candidate's Personal Record of Answers.

THIRD SIGNAL: End of test. Stop all work. If you finish sooner, raise your hand to signal the monitor.

TEST BOOKLET: This booklet contains *60* questions, all of equal weight. After the second signal, check your booklet to be sure it has all the questions and is not defective in any way. You are responsible for obtaining a complete booklet.

DIRECTIONS: Mark answers to all *60* questions on your Answer Sheet before the third signal. ONLY YOUR ANSWER SHEET WILL BE RATED. Use a soft No. 2 pencil to mark your answers. If you want to change an answer, erase it completely and then mark your new answer. For each question, pick the best answer. Then, on your Answer Sheet, in the row with the same number as the questions, blacken the circle with the same letter as your answer.

Here is a sample of how to mark your answers:

Question: When we add 5 and 3, we get:

(A) 2   (B) 4   (C) 8   (D) 9

19

Since the answer is 8, your Answer Sheet should be marked like this:

(A)   (B)   ●   (D)

WARNING: You are not allowed to copy answers from anyone or to use books, notes, or take the test for somebody else or to let somebody else take the test for you. You may be disqualified if you do not follow instructions.

This booklet is the property of the Department of Personnel of the City of New York and is worth more than $1,000. It must be returned with your Answer Sheet and scrap paper at the end of the test.

NOTE: Use of calculating devices is *not* permitted.

There will be no smoking anywhere in the building.

This is an unreleased test. Do not take this booklet with you when you leave.

ANSWER SHEET FOR MODEL EXAM I

1. Ⓐ Ⓑ Ⓒ Ⓓ
2. Ⓐ Ⓑ Ⓒ Ⓓ
3. Ⓐ Ⓑ Ⓒ Ⓓ
4. Ⓐ Ⓑ Ⓒ Ⓓ
5. Ⓐ Ⓑ Ⓒ Ⓓ
6. Ⓐ Ⓑ Ⓒ Ⓓ
7. Ⓐ Ⓑ Ⓒ Ⓓ
8. Ⓐ Ⓑ Ⓒ Ⓓ
9. Ⓐ Ⓑ Ⓒ Ⓓ
10. Ⓐ Ⓑ Ⓒ Ⓓ
11. Ⓐ Ⓑ Ⓒ Ⓓ
12. Ⓐ Ⓑ Ⓒ Ⓓ
13. Ⓐ Ⓑ Ⓒ Ⓓ
14. Ⓐ Ⓑ Ⓒ Ⓓ
15. Ⓐ Ⓑ Ⓒ Ⓓ

16. Ⓐ Ⓑ Ⓒ Ⓓ
17. Ⓐ Ⓑ Ⓒ Ⓓ
18. Ⓐ Ⓑ Ⓒ Ⓓ
19. Ⓐ Ⓑ Ⓒ Ⓓ
20. Ⓐ Ⓑ Ⓒ Ⓓ
21. Ⓐ Ⓑ Ⓒ Ⓓ
22. Ⓐ Ⓑ Ⓒ Ⓓ
23. Ⓐ Ⓑ Ⓒ Ⓓ
24. Ⓐ Ⓑ Ⓒ Ⓓ
25. Ⓐ Ⓑ Ⓒ Ⓓ
26. Ⓐ Ⓑ Ⓒ Ⓓ
27. Ⓐ Ⓑ Ⓒ Ⓓ
28. Ⓐ Ⓑ Ⓒ Ⓓ
29. Ⓐ Ⓑ Ⓒ Ⓓ
30. Ⓐ Ⓑ Ⓒ Ⓓ

31. Ⓐ Ⓑ Ⓒ Ⓓ
32. Ⓐ Ⓑ Ⓒ Ⓓ
33. Ⓐ Ⓑ Ⓒ Ⓓ
34. Ⓐ Ⓑ Ⓒ Ⓓ
35. Ⓐ Ⓑ Ⓒ Ⓓ
36. Ⓐ Ⓑ Ⓒ Ⓓ
37. Ⓐ Ⓑ Ⓒ Ⓓ
38. Ⓐ Ⓑ Ⓒ Ⓓ
39. Ⓐ Ⓑ Ⓒ Ⓓ
40. Ⓐ Ⓑ Ⓒ Ⓓ
41. Ⓐ Ⓑ Ⓒ Ⓓ
42. Ⓐ Ⓑ Ⓒ Ⓓ
43. Ⓐ Ⓑ Ⓒ Ⓓ
44. Ⓐ Ⓑ Ⓒ Ⓓ
45. Ⓐ Ⓑ Ⓒ Ⓓ

46. Ⓐ Ⓑ Ⓒ Ⓓ
47. Ⓐ Ⓑ Ⓒ Ⓓ
48. Ⓐ Ⓑ Ⓒ Ⓓ
49. Ⓐ Ⓑ Ⓒ Ⓓ
50. Ⓐ Ⓑ Ⓒ Ⓓ
51. Ⓐ Ⓑ Ⓒ Ⓓ
52. Ⓐ Ⓑ Ⓒ Ⓓ
53. Ⓐ Ⓑ Ⓒ Ⓓ
54. Ⓐ Ⓑ Ⓒ Ⓓ
55. Ⓐ Ⓑ Ⓒ Ⓓ
56. Ⓐ Ⓑ Ⓒ Ⓓ
57. Ⓐ Ⓑ Ⓒ Ⓓ
58. Ⓐ Ⓑ Ⓒ Ⓓ
59. Ⓐ Ⓑ Ⓒ Ⓓ
60. Ⓐ Ⓑ Ⓒ Ⓓ

Number Right _____

MODEL EXAM I

60 Questions—2½ Hours

Directions: Read each question and choose the best answer. On your answer sheet, darken the letter of the answer you choose.

1. Despite all efforts to avoid accidents and injuries, some will happen. If an employee is injured, no matter how small the injury, he or she should report it to the supervisor and have the injury treated. A small cut not attended to can easily become infected, and can cause more trouble than some injuries that at first seem more serious. It never pays to take chances.

   According to the preceding passage, an employee who gets a slight cut should

   (A) have it treated to help prevent infection
   (B) know that a slight cut becomes more easily infected than a big cut
   (C) pay no attention to it as it can't become serious
   (D) realize that it is more serious than any other type of injury

2.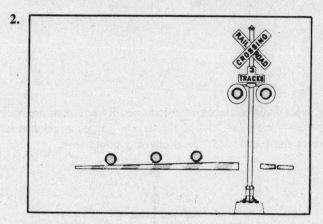

   If you come upon the scene above while driving along any roadway, you should

   (A) blow your horn
   (B) get out of your vehicle and move the barrier
   (C) come to a full stop and wait
   (D) stop, look, and proceed

3. A Sanitation Department rule states, "No passengers are to be carried in the cab of any sanitation vehicle at any time." As you are driving an empty pickup truck destined to collect bulky trash at a specific location, you pass your eight-year-old son on his way to school. You look at your watch and realize that the boy will be late to school. The best thing for you to do is

   (A) tell him to hop into the cab and you will drive him to school
   (B) have him stand on the running board and hold on tight to the window frame while you drive him to school

(C) tell him to climb into the cargo area for a ride

(D) tell him to hurry or he will be late to school

4. The sign beside a bridge reads, ''Maximum Safe Weight 6 Tons.'' Which of the following trucks may cross the bridge?

(A) A truck weighing 3 tons carrying a 4-ton load of gravel

(B) A truck weighing 5 tons returning empty after dumping its load

(C) A truck weighing 4 tons carrying 2½ tons of garden debris

(D) A truck weighing 3½ tons towing another truck weighing 3 tons

5. A car and a truck were in an accident. The inspector who tested the truck's brakes reported that they were *defective*. By this, he meant that the brakes

(A) were not working properly

(B) were in good shape

(C) had been relined

(D) had nothing to do with the accident

6.

The meaning of this sign is

(A) no parking

(B) no truck parking

(C) no trucks

(D) trucks only

7. As indicated by arrows on the street map shown below, Adams and River Streets are one-way going north. Main is one-way going south, and Market is one-way going northwest. Oak and Ash are one-way streets going east, and Elm is one-way going west.

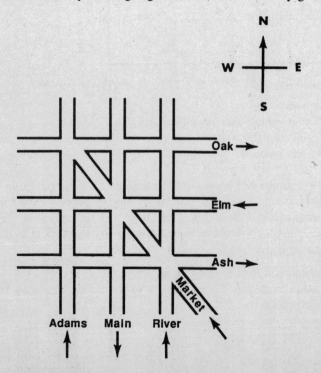

A collection truck which has just made a collection on River Street between Ash and Elm Streets must now make a special pickup at the intersection of Adams and Oak. In order to travel the shortest distance and not break any traffic regulations, the truck should turn

(A) left on Elm and right on Market
(B) left on Market and proceed directly to Oak
(C) left on Elm and right on Adams
(D) left on Oak and proceed directly to Adams

8. A sanitation worker who is loading a collection truck with trash and garbage in front of the apartment house at 3626 Broom Street, which is between Tremont Avenue and Walker Road, hears a crackling noise, looks overhead, and sees a sparking, dangling wire. The sanitation worker uses the building superintendent's phone to telephone the power company. Which of the following reports describes the situation most clearly and accurately?

(A) There is a dangerous live wire in the street on Broom Street hanging onto the garbage of an apartment house.
(B) A dangling wire is sparkling and crackling between Tremont Avenue and Walker Road.
(C) A live wire is hanging over the garbage on Broom Street right in front of my collection truck, and I can't collect the garbage.
(D) There is a crackling wire dangling over the sidewalk in front of 3626 Broom Street between Tremont Avenue and Walker Road.

9. Which of the objects below might be most dangerous to a sanitation worker?

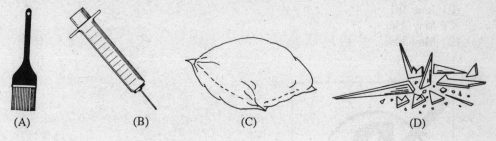

(A)   (B)   (C)   (D)

10. The examination announcement states, "You must have reached your 18th birthday by the date of appointment." This regulation means that

(A) if you are over 18 you should not apply
(B) you must file your application on your 18th birthday
(C) all sanitation workers are 18 years old when they begin work
(D) you will not be appointed if you are under 18

11. A driver about to cross a bridge sees these lane markings on the sign spanning the width of the bridge.

The driver

(A) may use the second lane from the right
(B) must use the far left lane
(C) may use the two middle lanes
(D) cannot cross the bridge

12. Sanitation Department policy states clearly that sanitation workers must not put their lives or safety in danger even if this means leaving refuse uncollected. The rules state that if a sanitation worker suspects that a receptacle contains an explosive, the worker should:

1. Go immediately to the nearest telephone and notify the police.
2. Move the collection vehicle at least 50 yards away from the suspicious receptacle.
3. Stand beside the vehicle and advise pedestrians not to enter the area.
4. Remain in the area to answer questions by the authorities.

Sanitation Worker Barbera approaches a wire basket with contents obscured by newspapers and hears the basket ticking. Barbera should

(A) move the newspapers to see what is inside the basket that is making it tick
(B) call a radio station so that a warning may be broadcast at once
(C) shout to pedestrians that they should step back 50 yards
(D) call the police

13. Garbage scows that break loose from their moorings and drift in the harbor present a hazard to navigation. For this reason, Department regulations require that all mooring ropes be visually inspected every 22 days. The ropes used to secure a barge were last inspected on May 3rd. The next scheduled inspection will be on

(A) May 23rd
(B) June 3rd
(C) May 25th
(D) May 30th

14.

TUES. & THURS.
10 A.M. - 2 P.M.

This sign means

(A) no parking except Tuesday and Thursday from 10 A.M. till 2 P.M.
(B) keep the street clear for street sweepers on Tuesday and Thursday between 10 A.M. and 2 P.M.
(C) brooms prohibited on Tuesday and Thursday from 10 P.M. to 2 A.M.
(D) parking for trucks only on Tuesday and Thursday from 10 A.M. to 2 P.M.

15. The gas gages below represent the readings on four Sanitation Department vehicles on Monday morning and again on Wednesday afternoon. Which vehicle used the most gasoline between Monday morning and Wednesday afternoon?

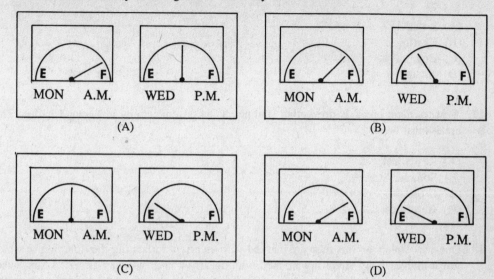

Answer questions 16 to 19 based on the street map shown below.

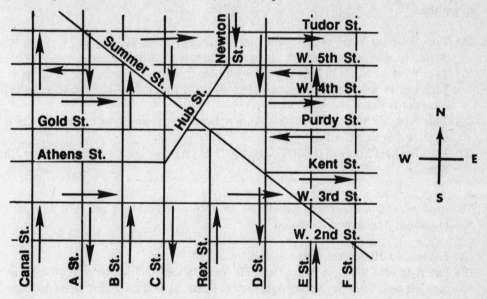

The city street map shown above is divided into four sanitation collection sectors as follows:

Sector Adam: Bounded by Tudor Street, Newton Street, Hub Street, Athens Street, and Canal Street.

Sector Boy: Bounded by Tudor Street, F Street, West 4th Street, Hub Street, and Newton Street.

Sector Charles: Bounded by West 4th Street, F Street, West 2nd Street, C Street, and Hub Street.

Sector David: Bounded by Athens Street, C Street, West 2nd Street, and Canal Street.

16. A sand spreader has gotten a flat tire and is disabled in the block bounded by West 4th Street, Summer Street, and Hub Street. The sand spreader is waiting for assistance in Sector

   (A) David
   (B) Charles
   (C) Boy
   (D) Adam

17. A water main break is preventing routine sanitation collections in Sector Charles. This break must be on

   (A) Rex Street
   (B) Summer Street
   (C) C Street
   (D) Hub Street

18. One-way streets on this map are marked with an arrow indicating the direction in which traffic moves. Any street not marked with an arrow is a two-way street. A supervisor who is in a car at the intersection of C Street and West 3rd Street needs to get to the corner of B Street and West 5th Street right away. What is the quickest and easiest way to get there?

   (A) Take West 3rd Street to E Street. Turn left onto E Street and proceed to West 5th Street. Turn left onto West 5th Street to the corner of B Street.
   (B) Take West 3rd Street to E Street. Turn left onto E Street and proceed to Kent Street. Turn left on Kent Street to Summer Street. Turn right onto Summer Street to the corner of B Street and West 5th Street.
   (C) Take West 3rd Street to Rex Street. Turn right onto Summer Street to the corner of B Street and West 5th Street.
   (D) Take West 3rd Street to B Street. Turn right onto B Street and go straight to West 5th Street.

19. The snow plow is proceeding in a northeast direction on Hub Street and must now plow Newton Street. The snowplow should

   (A) continue on Hub Street right into Newton Street
   (B) turn right onto West 5th Street, left onto Tudor Street, and left onto Newton Street
   (C) turn left onto West 5th Street, right onto C Street, right onto Tudor Street and right onto Newton Street
   (D) turn left onto West 5th Street, right onto B Street, right onto Tudor Street, and right onto Newton Street

20. A sanitation worker watched in horror as an automobile hit a woman who was crossing an avenue. When questioned by a police officer, the sanitation worker said that he was quite certain that the woman had *disregarded* the traffic light. The sanitation worker believed that the woman had

   (A) looked at the light
   (B) crossed with the light
   (C) paid no attention to the light
   (D) responded to the light

21. The Sanitation Department has strict rules against the collection of commercial refuse at public expense. If a sanitation worker suspects that commercial refuse has been put out alongside household trash, the procedure is as follows:

1. The worker must load only the household trash onto the truck.
2. The worker must report the location of the suspected violation and the identity of the suspected violator to the supervisor.
3. The supervisor must personally investigate the refuse which has been left at the curb.
4. The supervisor must advise the violator of the violation and of penalties if the violation is not corrected.

Sanitation Worker Planck is collecting refuse in front of a building which houses a butcher shop on the street level and small apartments upstairs. In the big bins out at the curb, Planck notices large bones and quantities of fat and other meat trimmings. The owner of the butcher shop is standing on the sidewalk in front of the shop. Planck should

(A) load everything onto the truck, then report the violation to the supervisor
(B) load whatever appears to be household trash and leave the refuse from the butcher shop
(C) tell the butcher to put the butcher shop refuse into the alley behind the shop for collection by private carters
(D) tell the butcher that all refuse must be properly bagged to be collected by the Sanitation Department

22.

The words on this sign mean

(A) dead end, no exit
(B) one way traffic
(C) no U turn, keep out
(D) special parking rules today; do not park here

23.

(a)   (b)   (c)   (d)

A sanitation worker must choose a pair of gloves from the shelf. Which gloves should the worker choose?

(A) (a) and (b)
(B) (a) and (d)
(C) (c) and (d)
(D) (b) and (c)

Answer questions 24 to 26 based on the chart below.

Vehicle ID	Mileage reading when taken out	Mileage reading when returned	Total miles driven
8723	12,562	12,591	
3406	16,086		37
8452		19,016	26

24. The total number of miles driven by vehicle 8723 was

(A) 29
(B) 31
(C) 33
(D) 39

25. The mileage reading when vehicle 3406 was returned was

(A) 16,049
(B) 16,113
(C) 16,123
(D) 16,129

26. The mileage reading when vehicle 8452 was taken out was

(A) 18,980
(B) 18,990
(C) 19,030
(D) 19,042

27. Which of the items below would be easiest to roll to the collection truck?

(a)

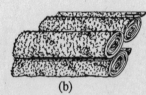
(b)

(c)

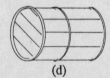
(d)

(A) (b)
(B) (c)
(C) (d)
(D) (a)

Use the calendar below to answer question 28.

```
┌─────────────────────┐
│      AUGUST         │
│  S  M  T  W  T  F  S│
│           1  2  3  4│
│  5  6  7  8  9 10 11│
│ 12 13 14 15 16 17 18│
│ 19 20 21 22 23 24 25│
│ 26 27 28 29 30 31   │
└─────────────────────┘
```

28. A sanitation worker was scheduled to work Tuesday through Saturday in the month of August. The worker reported regularly on every scheduled day. How many days did the worker work in August?

(A) 20 days
(B) 22 days
(C) 23 days
(D) 25 days

29. In rounding the corner from High Street to College Place, Sanitation Worker Tanaka sideswipes a car illegally parked beside a stop sign. Tanaka notes that the car is a black Pontiac Trans Am with New Jersey license plate 634 ATC. Which of the following statements reports this information most clearly and accurately?

(A) I hit a car at High Street and College Place. It was from New Jersey and it was illegal. It was black.
(B) A black Pontiac Trans Am was illegally parked at the corner of High Street and College Place, so I hit it.
(C) My truck hit an illegally parked black Trans Am, New Jersey 643 ATC on College Place at High Street.
(D) A black Pontiac Trans Am sportscar, New Jersey 634 ATC, was illegally parked near the corner of High Street and College Place when my truck hit it.

30. The rule concerning dead animals is as follows:

1. If the dead animal is too large to be removed by shovel, call your supervisor for instructions.
2. If the dead animal is wearing identifying tags, call your supervisor for instructions.
3. If the dead animal is not wearing identifying tags and is small enough to be removed by shovel, scoop it up by shovel and toss it into the hopper of the truck.
4. Operate the lift and dump of the truck even if the hopper is not yet filled.

Sanitation Worker Smythe finds a dead squirrel in the gutter near an accumulation of garbage bags set out by an apartment house. Smythe should

(A) check the animal's tags to identify its owner
(B) call the supervisor for instructions
(C) pick up the animal with a gloved hand and toss it into the hopper
(D) shovel up the animal, toss it into the hopper, and start the motor.

31.

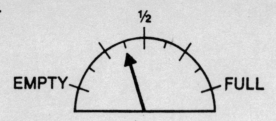

The arrow on the fuel gage above indicates that the gas tank is

(A) ⅓ full
(B) ⅜ full
(C) ¼ full
(D) ⅝ full

32.

The words on this sign mean that

(A) 500 people are working in the road
(B) for the next 1000 feet, people will be working in the road
(C) in 1000 feet, expect to find people working in the road
(D) please help the people working in the road for the next 1000 feet

33. A tug boat is capable of pulling a load weighing 12 tons. Two tug boats are attached to a garbage barge which weighs three tons and is currently loaded with 19 tons of garbage. How much more garbage can be loaded onto the barge before the tugs pull away from the dock?

(A) 2 tons
(B) 3 tons
(C) 12 tons
(D) 19 tons

34.

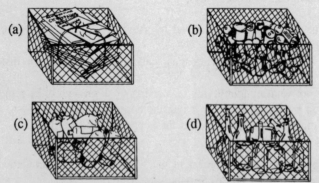

Into which of the recycling bins above should one place this object?

(A) bin (d)
(B) bin (a)
(C) bin (b)
(D) bin (c)

35. County recycling rules require that garbage and waste paper be packaged separately for collection. The regulations concerning violations of this rule are:

   1. First offense: Do the pickup and leave a polite note for the householder.
   2. Second offense: Do the pickup and leave an official warning for the householder.
   3. Third offense: Do not take the refuse; leave a summons.

   Sanitation Worker Lucas arrives at 72 Highland Way and discovers that the refuse at the curb includes a mingling of wet garbage, newspaper, other paper, and cans. Lucas leaves a note explaining the separation rules under the back door and collects the trash. Three days later Lucas arrives at 72 Highland Way and finds garbage in a plastic bag and newspapers stacked in paper bags. Lucas should

   (A) take the refuse but leave an official warning
   (B) leave the refuse without comment
   (C) take the refuse
   (D) ring the doorbell and hand the householder a summons

36. The snowstorm had just begun after a period of freezing rain, and the roadway was very slippery. Visibility was poor. Sanitation Worker Thompson, spreading a salt and sand mixture on Jones Street, suddenly skidded into a car driven by Bob Bozo. Three-year old Mary Bozo, who was standing on the front seat unrestrained, was thrown into the windshield and was injured. Thompson summoned Police Officer Haroldon who obtained an ambulance to take Mary Bozo to City Hospital. Which of the following reports should Thompson file to report the incident most clearly and accurately?

   (A) On a snowy day I hit a Bozo who was driving his three-year old without a car seat. The kid was hurt and Police Officer Haroldon sent her to City Hospital in an ambulance.
   (B) In the snowstorm I skidded into a car being driven by Bob Bozo on Jones Street. Mary Bozo, age three, was not in a car seat and was hurt in the accident. Police Officer Haroldon arranged her transportation to City Hospital.
   (C) Mary Bozo was hurt when I hit a car being driven by Bob Bozo in the snow on Jones Street. She went to the hospital in an ambulance not in a car seat.
   (D) On Jones Street which I was sanding in the snow, I hit Bob Bozo and Police Officer Haroldon took Mary to City Hospital. Mary is three.

37. Street-sweeping machines work in a dusty environment and should have their oil changed at least every 2000 miles. Sweeper #Y-98 last had its oil changed at 5786 miles. The odometer (mileage gage) now reads 7739. How many more miles may the sweeper be driven before it should be brought in for an oil change?

   (A) 37 miles
   (B) 45 miles
   (C) 47 miles
   (D) 49 miles

38.

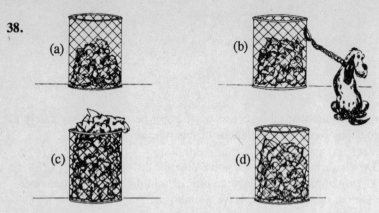

Which of the above baskets will be most difficult to empty?

(A) (c)
(B) (d)
(C) (a)
(D) (b)

39.

On the basis of the traffic signal above, it would be correct to say that

(A) drivers approaching from the right cannot see any traffic signals
(B) drivers approaching the signal straight on see a green arrow pointing to the left
(C) drivers approaching from the left are guided by five different traffic signals
(D) drivers approaching straight on see a red arrow pointing to the right

40. There is a series of entries in Sanitation Worker Falcone's personnel file. In which order did these events probably occur?

1. The doctor applied antiseptic and bandaged the cut.
2. Falcone called the supervisor to say that he would be out for the rest of the week.
3. Falcone slipped from the back of the collection truck and gashed his leg on the side of the hopper.
4. Falcone went home and put his leg up.

(A) 3-2-1-4
(B) 4-3-1-2
(C) 3-1-4-2
(D) 2-1-4-3

41. The rules state that sanitation workers are not to search through the refuse and must not remove any items that appear to be of interest to them. The requirement is that all refuse be dumped into the hopper immediately, and if a container is too heavy or too bulky for one person, two workers may lift together. In general, there are to be no exceptions. In which of the following situations should a sanitation worker exercise independent judgment and disobey this regulation?

(A) Four kitchen chairs in excellent condition have been put at curbside, and the sanitation worker needs new chairs badly.
(B) A large, bulky rug has a very bad odor.
(C) A can full of ashes is backbreakingly heavy.
(D) A baby's cries can be heard coming from a closed bag.

42. If a sign reads "Bridge Freezes Before Roadway," it means

(A) bridge may be icy
(B) detour—spilled ice on bridge
(C) watch for frozen roads
(D) cold weather forecast for tonight

43. A sanitation worker's regular work week is 40 hours, 8 hours a day for 5 days. If a worker is required to work more than 40 hours in any one work week, the worker is paid overtime for each additional hour worked. Sanitation Worker McVoy is scheduled to work a regular collection run from 7 A.M. to 3 P.M. on Monday through Friday. On Tuesday afternoon it begins snowing, and McVoy is called back to drive the sander from 6 P.M. to 10 P.M. On Thursday, McVoy is asked to remain after the regular collection run to help with snow removal and works until 8:30 P.M. For how many hours of overtime will McVoy be paid this week?

(A) 8½ hours
(B) 9 hours
(C) 9½ hours
(D) 10½ hours

44.

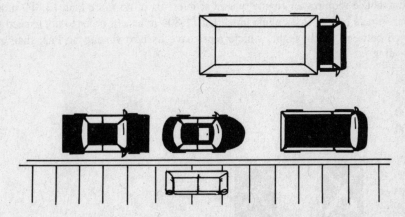

In the scene above, which would be the best, most efficient way to get the sofa to the truck without touching any of the vehicles?

(A) over the sports car
(B) around the back of the second car
(C) between the two cars
(D) between the van and the sports car

45. The Sanitation Department has a carefully organized time schedule, and if a crew member is late to work the whole schedule is thrown off. Lateness is a serious offense and frequent lateness is cause for disciplinary action. Public transportation problems which are reported over the news and major highway tie-ups are acceptable excuses. Car troubles, late baby-sitters, and other personal excuses will be dealt with on a case-by-case basis. A sanitation worker who arrives late for work must:

1. Report in to the supervisor
2. Join the assigned crew and get to work
3. Return to the supervisor at the end of the work day to fill out form #P-546

Sanitation Worker Vorperian arrives at the subway station and learns that there has been a fire in the tunnel and no subway trains have come through for over an hour. However, Vorperian has been standing only five minutes when a train arrives. She arrives at work about eight minutes late. Vorperian should

(A) report right to work; eight minutes is not worth worrying about
(B) report to the supervisor and hold up the crew a bit longer
(C) fill out form #P-546
(D) ignore the incident; it was an excused lateness

46. While Sanitation Worker Park was helping to shift the load on a garbage barge, a high wind came up and the barge began to roll heavily on the waves. Park stepped into a slippery patch just as the barge pitched and fell headlong into the garbage. In a report of the incident, Park wrote that the high seas *contributed* to the accident. Park meant that the wind

(A) caused the accident
(B) was partly responsible for the accident
(C) was the only explanation for the accident
(D) had nothing to do with the accident

47. A sanitation vehicle requires an engine tuneup at intervals of no more than 12,000 miles. A particular vehicle had its first engine tuneup at 11,804 miles. In order to not exceed the time allowed between tuneups, the vehicle must have its next tuneup no later than at a mileage reading

(A) 23,608
(B) 23,804
(C) 24,000
(D) 24,800

48.

A driver who sees the above sign on a bridge knows that

(A) she should use the left lane
(B) the two right lanes are one-way streets
(C) the left lane is one-way in the direction opposite to that in which she is proceeding
(D) there has been an accident in the two right lanes

49. Which of the implements below would be most appropriate for picking up this pile of dirt?

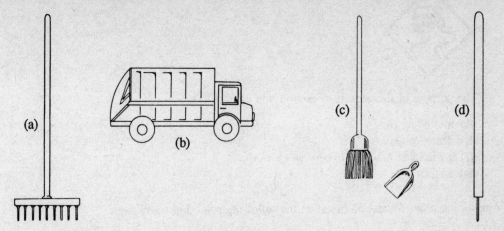

Answer questions 50 to 52 based on the duty chart below.

7 to 3 SHIFT

Sun.	Mon.	Tues.	Wed.	Thurs.	Fri.	Sat.
	Patel	Patel		Patel	Patel	Patel
Gomez	Gomez		Gomez	Gomez	Gomez	
Sobel			Sobel	Sobel	Sobel	Sobel
	Quirk		Quirk	Quirk	Quirk	Quirk

50. The sanitation worker who works the most consecutive days is

(A) Patel
(B) Quirk
(C) Sobel
(D) Gomez

51. When Quirk gets home on Saturday afternoon, he knows that he is off until

(A) Monday afternoon
(B) Sunday afternoon
(C) Wednesday morning
(D) Monday morning

52. The day of the week on which one of these workers will go out with an entirely unfamiliar crew is

(A) Sunday
(B) Tuesday
(C) Wednesday
(D) Thursday

53.

The purpose of this sign is to caution you against

(A) a winding road
(B) drunk drivers
(C) a road that may be slippery when wet
(D) a steep hill

Answer questions 54 and 55 based on the following rules and traffic sign.

Drivers of Sanitation Department street-sweeping machines are assigned specific routes. Traffic signs are posted so that these streets will be clear of vehicles at street-cleaning times. Department of Parking Violations tow trucks have the duty of towing away illegally parked vehicles that interfere with street cleaning. "Parking" is defined as stopping a vehicle at the curb and leaving the vehicle. "Standing" is defined as stopping a vehicle at the curb with the driver remaining in or next to the vehicle. Whenever parking is prohibited, standing is prohibited as well. Sanitation workers are not permitted to touch cars but must use reasonable means to clear the streets for their scheduled cleanings. The following sign is posted on Michigan Avenue:

```
┌─────────────────────────┐
│                         │
│       NO PARKING        │
│     10 A.M. to 1 P.M.   │
│      Mon., Wed., Fri.   │
│    ─────────────────    │
│                         │
│    TOW-AWAY ZONE        │
│                         │
└─────────────────────────┘
```

54. Sanitation Worker Canellos has been assigned to clean Michigan Avenue on Wednesday at 11:30 A.M. In the middle of the block, Canellos comes upon a motorcycle parked alongside the curb. Canellos should

(A) blow the horn of the sweeper in hopes that the owner of the motorcycle is not far away
(B) lift the motorcycle onto the sidewalk
(C) call the supervisor for instructions
(D) summon a Department of Parking Violations tow truck

55. On Friday at 12:15, Sanitation Worker Ahmet finds a panel truck at the curb along Michigan Avenue. The driver of the panel truck is standing alongside the truck talking to a pedestrian who is walking a large dog. Ahmet should

(A) summon a Department of Parking Violations tow truck
(B) ask the driver to move the panel truck
(C) pat the dog
(D) ask the pedestrian to move along

56. When moving a sanitation vehicle in reverse, the driver must do the following in the following order:

1. Turn on the flashing blue light
2. Activate the high-pitched beeper
3. Check both side mirrors
4. Stick his or her head out the window and look both ways
5. Blow the horn.
6. When safe, put the vehicle into reverse and back up.

Sanitation Driver Djilani must get a large collection vehicle out of a narrow dead-end street. Djilani has activated the flashing blue light and the high-pitched beeper and has checked the mirror on the driver's side. The next thing for Djilani to do is to

(A) look both ways out the window
(B) shout at the children in the street to get out of the way
(C) check the mirror on the passenger side
(D) back up

57. Sanitation workers face a danger to their health when handling infectious waste. Infectious waste must be packaged in protective containers and must be clearly labelled. Sanitation worker Rapski arrives to make streetside collection in front of 1101 Bramson Boulevard, a large building with doctors' offices on the ground level. Some of the plastic bags have burst open, and squares of bloody gauze litter the area. None of the bags or packages bear warning labels. Rapski leaves the refuse uncollected and reports to the supervisor. Which report describes the situation most clearly and accurately?

(A) There is uncontainerized medical waste on the sidewalk in front of 1101 Bramson Boulevard.
(B) Toxic waste is all over the place in front of 1101 Bramson Boulevard. I won't pick it up because it is dangerous.
(C) The infectious waste at 1101 Bramson Boulevard is not packaged or labelled. It is making a mess.
(D) This is to report unidentified infectious waste 1011 Bramson Boulevard. It needs to be wrapped and labelled because it is all bloody.

58. Incinerating household refuse releases smoke and soot particles and contributes to air pollution. For this reason, large apartment houses are no longer permitted to use their incinerators. These large buildings have installed compactors which compress a great deal of waste into a very small space. Smaller buildings and private houses put out their waste in heavy-gauge plastic bags. All buildings cooperate by putting cans and glass bottles into separate metal garbage cans. A collection truck arriving to collect refuse from a very large apartment house can expect to find

(A) compacted cans and heavy-gauge plastic bags
(B) cans full of ashes and separate cans of glass jars and bottles
(C) containers of smoke and soot, glass, paper and plastic bags
(D) very heavy packages of refuse and cans of metal and glass

59. The driver of the snow plow is sent out during a snowstorm that is so heavy that street signs cannot be read. The route is described in terms of left and right turns. The instructions read: Proceed straight on the road out of the garage to the first cross street; turn left on that street, then take the first right; proceed to the next intersection and turn right again;

take the next two left turns in order and then turn right. At this point in the route, the snowplow is

(A) back where it began
(B) facing in the same direction as when it started out
(C) facing in the direction from which it came, as if to return to the garage
(D) in a dead end

60. Sanitation workers must be in good physical condition at all times. If a sanitation worker feels ill, the worker may become inattentive and suffer an accident on the job. Sanitation workers, therefore, are not discouraged from using their sick leave. The rule governing paid sick leave is that the worker call the supervisor at least two hours before he or she is scheduled to report to work, except in case of emergency. If the worker is not truly ill or if notification is late, the worker will not be paid for the absence. Which one of these workers will be paid for a sick day?

(A) Sanitation Worker Silver who is scheduled to report at 3 P.M. and who calls in at noon to say that his mother-in-law has a headache.
(B) Sanitation Worker Hawkins who is scheduled to report at 7 A.M. calls in at 5:30 A.M. to say that he has a hangover.
(C) Sanitation Worker Miceli who is scheduled to report at 7 A.M. calls at 6:15 A.M. to say that she fell down the stairs on her way to work and is now at the Emergency Room of Central Hospital having a broken finger set.
(D) Sanitation Worker Green who is scheduled to report at 9 A.M. calls at 7 A.M. to report that the dog is vomiting and must be taken to the vet.

MODEL EXAM I ANSWER KEY

1. A	11. A	21. B	31. B	41. D	51. D
2. C	12. D	22. B	32. C	42. A	52. B
3. D	13. C	23. B	33. A	43. C	53. C
4. B	14. B	24. A	34. A	44. D	54. D
5. A	15. D	25. C	35. C	45. B	55. B
6. C	16. D	26. B	36. B	46. B	56. C
7. A	17. A	27. C	37. C	47. B	57. A
8. D	18. C	28. C	38. D	48. A	58. D
9. B	19. D	29. D	39. B	49. C	59. B
10. D	20. C	30. D	40. C	50. C	60. C

EXPLANATIONS

1. **(A)** The paragraph says that even small cuts must be treated because they, too, may become infected. It does *not* say that small cuts are more easily infected than larger ones, so choice (B) is incorrect. Choice (C) contradicts the whole paragraph. Choice (D) is wrong because the paragraph does not say that all small cuts are more serious than other injuries but only that they might (if unattended) cause more trouble than more serious injuries that are treated from the start.

2. **(C)** The picture shows a railroad crossing with the barrier arm down signifying that a train is approaching. A driver must stop and wait until the train passes and the arm is raised signifying that it is safe to cross the track.

3. **(D)** Choice (A) is exactly in opposition to the rule. You must not disobey rules, even for your own child. Choices (B) and (C), while not prohibited by the stated rule, represent outrageously unsafe practices, especially with regard to an eight-year-old.

4. **(B)** Since the truck weighs a total of only 5 tons, it may cross safely. The weight of the truck in (A) is 7 tons; the weight of the truck in (C) is 6½ tons; the total of the two attached trucks which might well be on the bridge at the same time in (D) is 6½ tons.

5. **(A)** *Defective* means imperfect or faulty. Defective brakes do not work properly.

6. **(C)** A line slashed across a picture means that whatever is represented by the picture is prohibited. In this instance, trucks are not allowed. If the sign were to mean "No truck parking," there would be a large "P" superimposed on the truck. A slash through a "P" means "No parking."

7. **(A)** The map on the next page shows the truck beginning on River between Ash and Elm and indicates the route, left on Elm and right on Market, to the special pickup at the intersection of Oak and Adams at the X. Choice (C) would be slightly longer. Both (B) and (D) are impossible from this starting point and (D), in addition, would involve going the wrong way on a one-way street.

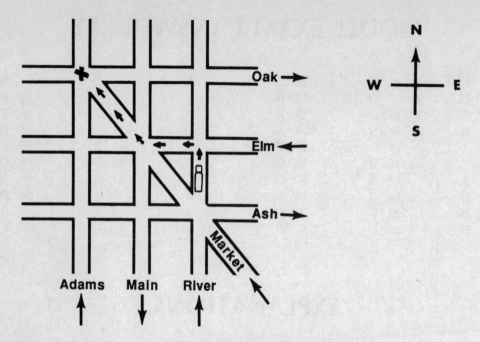

8. **(D)** The most important information here is the precise location of the downed live wire so that the power company can come to the site at once. The street name and number pinpoint the location exactly, and naming the cross streets helps the power authority to plan the best route.

9. **(B)** The hypodermic needle is most likely to present a danger of infectious disease, including AIDS. The broken glass is dangerous, but the needle presents greater danger. Paintbrushes and pillows are not dangerous.

10. **(D)** Read carefully. You must be 18 to be appointed. You may apply before you are 18, but you cannot be appointed until you turn 18. You may be older than 18 when you apply or when you start work.

11. **(A)** The way to read these signs is "X means do not enter," and "arrow means OK to enter." There is an arrow over the second lane from the right, so any driver may use that lane. The far left lane also has an arrow, so a driver may use that lane, but, since there are three lanes with arrows, the driver is not required to use the far left lane. The driver may use any lane with an arrow. The driver cannot use the two middle lanes because those lanes are marked with X's.

12. **(D)** The rule is very clear. Barbera must call the police. Most certainly Barbera should not disturb the contents of the ticking basket because the motion might set off a bomb.

13. **(C)** May 25th is 22 days after May 3rd. 3 + 22 = 25.

14. **(B)** The slash through the "P" means "No parking." The broom tells the reason that parking is not permitted: for street cleaning. The times indicated on the sign represent the times that parking is prohibited. In this case, "No parking on Tuesday and Thursday from 10 A.M. to 2 P.M. for street cleaning" means the same as "Keep the street clear for street sweepers on Tuesday and Thursday between 10 A.M. and 2 P.M."

15. **(D)** The gage on vehicle (D) went from almost full to almost empty. That vehicle used nearly a whole tankful of gasoline. None of the others used much more than half a tank.

16. **(D)** Sector Adam is bounded by heavy black lines. The disabled spreader is at X.

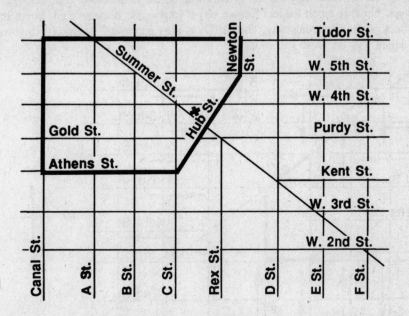

17. **(A)** Sector Charles is bounded by heavy black lines on the map below. Only Rex Street is entirely within Sector Charles. All the other choices involve streets which border on or go through at least one other sector. Since the question reads, ". . .the break *must* be on. . ." the answer must be Rex Street.

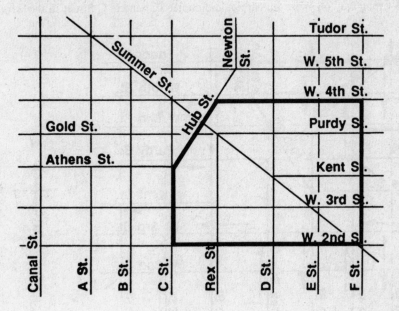

18. **(C)** The shortest and easiest way to get from the supervisor's starting point at "*" to the destination at "X" is marked by the heavy dark line on the map below. Choice (A) is a possible route, but it is much longer. Choice (B) is impossible because Kent Street is a one-way street going the wrong way; choice (D) is also impossible because it involves going the wrong way on West 3rd Street.

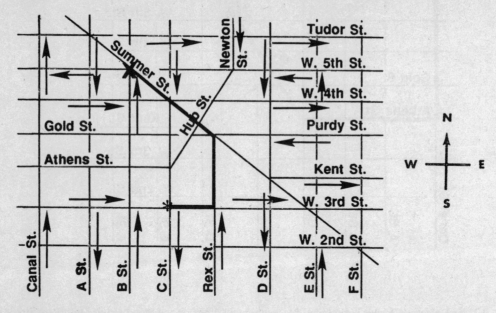

19. **(D)** The heavy dark line on the map below, from one dot to the other, marks a long, roundabout route. This route is required because anything shorter would require going the wrong way on a one-way street. Choice (A) enters Newton Street the wrong way; choice (B) goes the wrong way on West 5th Street; and choice (C) enters C Street in the wrong direction.

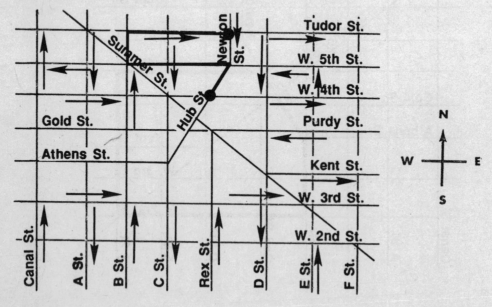

20. **(C)** To *disregard* is to ignore, to neglect, or to pay no attention to. The sanitation worker's observation was that the woman had stepped off the curb and started across the street without looking.

21. **(B)** Rule 1 says that the worker should load only the household trash. Then it is up to the supervisor to determine whether or not the remaining waste material is commercial refuse and to act accordingly. The sanitation worker should not make the decision as to violation, but should simply follow the rules.

22. **(B)** A sign prohibiting entry prohibits entry to a one-way street (going the other way) or to a roadway on which traffic is never permitted (like a pedestrian mall). A dead-end street must be a two-way street; it might have a sign warning of no exit but it would never prohibit entry.

23. **(B)** Glove (a) is the only work glove for the right hand. It must be paired with a work glove for the left hand. Both (c) and (d) are work gloves for the left hand, but only choice (B) offers one of these paired with (a). Glove (b) is for the right hand but it is not appropriate for a sanitation worker.

24. **(A)** Answer this question by subtracting the mileage reading when the vehicle was taken out from the mileage reading when it was returned. This difference represents the total number of miles driven. $12,591 - 12,562 = 29$

25. **(C)** Add the total number of miles driven to the mileage reading when the vehicle was taken out to find the mileage reading when the vehicle was returned. $16,086 + 37 = 16,123$

26. **(B)** Subtract the total number of miles driven from the mileage reading when the vehicle was returned to find out what the mileage reading was when the vehicle was taken out. $19,016 - 26 = 18,990$

The completed chart should look like this:

Vehicle ID	Mileage reading when taken out	Mileage reading when returned	Total miles driven
8723	12,562	12,591	29
3406	16,086	16,123	37
8452	18,990	19,016	26

27. **(C)** This is an easy question for which to choose the answer. Obviously, a barrel is easier to roll than a chair, a bulky pile of carpeting, or an irregularly shaped garbage bag. The difficulty with this question is marking the correct answer space. You must pay attention to what you are doing. The letter of the object that is easiest to roll is (d), but the letter of the correct answer is (C). You *must* mark the letter of the correct answer choice to get credit for a correct answer.

28. **(C)** There is more than one way in which you can come up with the correct answer. You could just count the days. Or you could see that August has 31 days of which 8 are not Tuesday, Wednesday, Thursday, Friday, or Saturday. Then you could subtract 8 from 31 $(31 - 8 = 23)$ to find out that the worker worked on 23 days.

29. **(D)** This statement accurately identifies the vehicle and the location of the accident and briefly tells the circumstances. Choice (C) gets the license plate number wrong. Choices (A) and (B) are incomplete.

30. **(D)** Start with common sense. Did you ever see a squirrel wearing identifying tags? Eliminate choice (A). The rules concerning small dead animals are very clear, and this

is a small animal. There is no need to call the supervisor. Eliminate choice (B). The rules mention use of a shovel two times. There should be no question that dead animals are not to be handled, even with gloves. Choice (C) is wrong. That leaves (D) as the correct answer.

31. **(B)** There are eight equally spaced markings on the gage, so you know it is divided into eighths. Full would be 8 eighths or $\frac{8}{8}$. Count back from full and find that $\frac{5}{8}$ of the of the fuel has been used. $\frac{8}{8} - \frac{5}{8} = \frac{3}{8}$. What is left is $\frac{3}{8}$ of a tank, so the tank is $\frac{3}{8}$ full. Be careful in choosing your answer. The indicator is $\frac{5}{8}$ *from* full; this does *not* mean that the tank is $\frac{5}{8}$ full.

32. **(C)** Consider this the 1000 foot warning. The sign cautions you that you can expect construction ahead in 1000 feet, and you should stay alert. A sign that means road work will begin at that spot and continue for 1000 feet will read "Road work next 1000 feet."

33. **(A)** The two tugs together can pull a total weight of 24 pounds. $12 + 12 = 24$. The current weight of the barge plus its load is 22 tons. 3 T (weight of the barge) + 19 T (current load) = 22 T. $24 - 22 = 2$ tons can be added to the load.

34. **(A)** The glass jug goes into the recycling bin for glass bottles. The bin for glass bottles is labelled (d). The answer to the question is (A). Watch out!

35. **(C)** The note that Lucas left after the first offense did its job. The householder has separated the garbage; there is no second offense. Lucas can pick up all the trash and continue the route.

36. **(B)** This statement gives all the information in a clear, direct way. (A) tells what Thompson thought, but it leaves out names and location. (D) is almost accurate, but sends Mary to the hospital in the police car rather than by ambulance.

37. **(C)** First you must find out how many miles the sweeper has travelled since its last oil change. Subtract the reading at the last oil change from the current reading. $7739 - 5786 = 1953$. Then subtract the number of miles the sweeper has travelled since its last oil change from the 2000 miles it could possibly go till its next oil change to find out how many more miles it can still go. $2000 - 1953 = 47$.

38. **(D)** Regardless of how much trash there is in any of the cans, the one that presents a real problem in terms of lift and dump is the one with the dog chained to it. Just be sure to get the letter of the answer right in marking the answer sheet.

39. **(B)** Drivers approaching from the right and the left each see three signals. You are looking at the signal straight on. The arrow points to the left; the letter under the arrow says that the arrow is green.

40. **(C)** The series of entries is in a personnel file, so clearly they concern an incident at work. Logically, then, we must start with the accident, item 3. This narrows to choices (A) and (C). It is reasonable to assume that Falcone got medical attention and was told to stay off the leg before calling the supervisor with a time period.

41. **(D)** Common sense! Don't throw out the baby with the trash.

42. **(A)** Choices (B), and (D) make no sense as meanings for a road sign. Choice (C) might be possible, but the sign refers specifically to bridges and suggests that bridges freeze before

roads. If one has already skidded on icy bridges, one should not need to be cautioned about roads. But, if the road is not icy and the temperature is low or falling, a warning about the possibility of an icy bridge is very much in order.

43. **(C)** On Tuesday McVoy pluts in an extra 4 hours from 6 P.M. to 10 P.M. $10 - 6 = 4$. On Thursday, McVoy remains from the regularly scheduled quitting time of 3 P.M. until 8:30 P.M. From 3 P.M. to 8:30 P.M. is 5½ hours. 8:30 is half past eight. $8½ - 3 = 5½$. Combine the two periods of overtime. $4 + 5½ = 9½$ total hours of overtime.

44. **(D)** Over the sports car might be the shortest route, but there is a real danger of dropping the sofa on the car. Between the two cars the space is too tight. Around the back of the second car is all right, but it is the long way round. The space between the van and the sports car is adequate and is not too far from the back of the truck.

45. **(B)** The rule is that *any* lateness must be reported to the supervisor. Eight minutes is not long, but nothing in the rules as given here exempts an eight-minute lateness from the requirement. Vorperian need not worry about discipline. The form #P-546 will have a space in which to record the documented subway problem which will excuse the lateness.

46. **(B)** To *contribute* is to have a share in the bringing about of an event or to be partly responsible for it. It was the combination of slippery garbage and the pitching of the barge caused by the wind that together made Park fall.

47. **(B)** A Sanitation vehicle must have an engine tuneup by every 12,000 miles, but it may have the tuneup earlier if convenient. We know this because we are told that this vehicle had its first tuneup after only 11,804 miles. Adding 12,000 miles to the mileage reading at the first tuneup, we learn that the vehicle must have its next tuneup by the time it has travelled a total of 23,804 miles. $11,804 + 12,000 = 23,804$. It may have a tuneup sooner, but it may not have a tuneup later.

48. **(A)** The Xs over the two right lanes indicate that those lanes are closed to traffic in this direction. It could be that there is an accident in those lanes; it could be that traffic moves in the other direction; or there could be some other reason that the lanes are closed. The lane with the arrow is open to traffic approaching. In this question the lane with the arrow is on the left.

49. **(C)** The appropriate implements for picking up a small pile of dirt are dustpan and broom. Using a garbage truck or backhoe would be overkill. The spiked stick is appropriate for picking up papers and a few odd leaves, but it would not be efficient for picking up dirt.

50. **(C)** Sobel works Wednesday, Thursday, Friday, Saturday, and Sunday—five consecutive days. Quirk works four consecutive days; Patel and Gomez work only three days in a row.

51. **(D)** Quirk has Sunday off. He works the 7:00 A.M. to 3:00 P.M. shift. When he finishes work on Saturday afternoon, he is off until Monday morning.

52. **(B)** Of the four workers on this schedule, only Patel is scheduled to work on Tuesday. Patel will not work alone but will be joined by a different crew.

53. **(C)** The picture on the sign represents a car skidding and fishtailing on a slippery road. Drunk drivers should not be on the road at all. The sign warning of a winding road looks like this:

This is the sign warning of a steep hill ahead:

54. **(D)** Sanitation workers are not allowed to touch vehicles, so Canellos may not lift the motorcycle out of the way even if he could physically do so. Blowing the horn is not mentioned as part of the procedure and could be annoying to residents of the block. The instructions are clear; send for a tow truck. There is no need to involve a supervisor.

55. **(B)** The distinction is made between parking and standing precisely because the driver of a standing vehicle is available to move it immediately. Ahmet should first ask the driver to move the truck. If the driver refuses, then Ahmet must get a tow truck so that the street can be cleaned.

56. **(C)** Just follow the rules step-by-step. Djilani has only checked one mirror. The next thing to do is to check the other mirror.

57. **(A)** This is the only statement based strictly on the facts. Rapski cannot know that the waste is toxic or infectious, but it most certainly is medical waste. Statement (D), in addition, has the building number wrong.

58. **(D)** If a compactor compresses a great deal of waste into a very small space, it stands to reason that the package will be very heavy. (B) is incorrect because with burning prohibited, there are no ashes to put out in the cans. Reread if necessary. The paragraph speaks of large apartment buildings, smaller buildings and private houses, but the question asks specifically about large apartment houses.

59. **(B)** The best way to answer this question is to make your own diagram following all the turns. Use scratch paper, if permitted, at your exam. Otherwise use margins of your test booklet. Remember to mark your answer on the answer sheet.

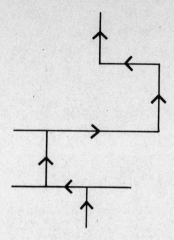

60. **(C)** Miceli gave only 45 minutes notice, but she could hardly have been expected to call in sick before the injury happened. Under these emergency circumstances, her report is as timely as possible. There is no doubt that this is a personal sickness situation for which she should be paid sick leave. In (A) Silver called in with time to spare, but he is not ill. In (B), Hawkins' hangover would certainly render unfit him for duty, but he called in too late. Under the rules, he will not be paid. In (D) Green gave just adequate notice, but the dog's illness is not a personal illness which would interfere with his work performance.

ANSWER SHEET FOR MODEL EXAM II

1. Ⓐ Ⓑ Ⓒ Ⓓ   16. Ⓐ Ⓑ Ⓒ Ⓓ   31. Ⓐ Ⓑ Ⓒ Ⓓ   46. Ⓐ Ⓑ Ⓒ Ⓓ

2. Ⓐ Ⓑ Ⓒ Ⓓ   17. Ⓐ Ⓑ Ⓒ Ⓓ   32. Ⓐ Ⓑ Ⓒ Ⓓ   47. Ⓐ Ⓑ Ⓒ Ⓓ

3. Ⓐ Ⓑ Ⓒ Ⓓ   18. Ⓐ Ⓑ Ⓒ Ⓓ   33. Ⓐ Ⓑ Ⓒ Ⓓ   48. Ⓐ Ⓑ Ⓒ Ⓓ

4. Ⓐ Ⓑ Ⓒ Ⓓ   19. Ⓐ Ⓑ Ⓒ Ⓓ   34. Ⓐ Ⓑ Ⓒ Ⓓ   49. Ⓐ Ⓑ Ⓒ Ⓓ

5. Ⓐ Ⓑ Ⓒ Ⓓ   20. Ⓐ Ⓑ Ⓒ Ⓓ   35. Ⓐ Ⓑ Ⓒ Ⓓ   50. Ⓐ Ⓑ Ⓒ Ⓓ

6. Ⓐ Ⓑ Ⓒ Ⓓ   21. Ⓐ Ⓑ Ⓒ Ⓓ   36. Ⓐ Ⓑ Ⓒ Ⓓ   51. Ⓐ Ⓑ Ⓒ Ⓓ

7. Ⓐ Ⓑ Ⓒ Ⓓ   22. Ⓐ Ⓑ Ⓒ Ⓓ   37. Ⓐ Ⓑ Ⓒ Ⓓ   52. Ⓐ Ⓑ Ⓒ Ⓓ

8. Ⓐ Ⓑ Ⓒ Ⓓ   23. Ⓐ Ⓑ Ⓒ Ⓓ   38. Ⓐ Ⓑ Ⓒ Ⓓ   53. Ⓐ Ⓑ Ⓒ Ⓓ

9. Ⓐ Ⓑ Ⓒ Ⓓ   24. Ⓐ Ⓑ Ⓒ Ⓓ   39. Ⓐ Ⓑ Ⓒ Ⓓ   54. Ⓐ Ⓑ Ⓒ Ⓓ

10. Ⓐ Ⓑ Ⓒ Ⓓ   25. Ⓐ Ⓑ Ⓒ Ⓓ   40. Ⓐ Ⓑ Ⓒ Ⓓ   55. Ⓐ Ⓑ Ⓒ Ⓓ

11. Ⓐ Ⓑ Ⓒ Ⓓ   26. Ⓐ Ⓑ Ⓒ Ⓓ   41. Ⓐ Ⓑ Ⓒ Ⓓ   56. Ⓐ Ⓑ Ⓒ Ⓓ

12. Ⓐ Ⓑ Ⓒ Ⓓ   27. Ⓐ Ⓑ Ⓒ Ⓓ   42. Ⓐ Ⓑ Ⓒ Ⓓ   57. Ⓐ Ⓑ Ⓒ Ⓓ

13. Ⓐ Ⓑ Ⓒ Ⓓ   28. Ⓐ Ⓑ Ⓒ Ⓓ   43. Ⓐ Ⓑ Ⓒ Ⓓ   58. Ⓐ Ⓑ Ⓒ Ⓓ

14. Ⓐ Ⓑ Ⓒ Ⓓ   29. Ⓐ Ⓑ Ⓒ Ⓓ   44. Ⓐ Ⓑ Ⓒ Ⓓ   59. Ⓐ Ⓑ Ⓒ Ⓓ

15. Ⓐ Ⓑ Ⓒ Ⓓ   30. Ⓐ Ⓑ Ⓒ Ⓓ   45. Ⓐ Ⓑ Ⓒ Ⓓ   60. Ⓐ Ⓑ Ⓒ Ⓓ

Number Right _____

MODEL EXAM II

60 Questions—2½ Hours

Directions: Read each question and choose the best answer. On your answer sheet, darken the letter of the answer you choose.

1. Sanitation workers assigned to operate street sweeping machines are given route sheets. The route sheet is used to make a record of the streets which have been cleaned. This information is used by others to keep the Sanitation Department's Street Sweeping Report.

   Sanitation workers who operate street sweepers

   (A) make a record of streets that they have cleaned
   (B) spend most of their time recording route sheet information
   (C) sweep the streets in any order
   (D) fill out the Sanitation Department's Street Sweeping Report

2. Department of Sanitation collection trucks have air brakes. The air brakes are used to slow down or stop a truck. After starting the engine, the driver must check the air pressure by looking at the air pressure gage. Within five minutes the air pressure gage should read 110 pounds. The truck should not be moved until the gage reads at least 110 pounds.

   Sanitation trucks should not be moved until the air pressure gage reads

   (A) 5 pounds
   (B) 50 pounds
   (C) 90 pounds
   (D) 110 pounds

Answer questions 3 and 4 only on the basis of the following information about speed limits for sanitation trucks: The following maximum speed limits have been established for safe operation of department trucks:

   1. 15 MPH on all paved streets within the city limits
   2. 40 MPH on all expressways
   3. 10 MPH on all dirt roads

3. The maximum speed that a truck should go on dirt roads within the city limits is

   (A) 15 MPH
   (B) 40 MPH
   (C) 10 MPH
   (D) 25 MPH

4. The maximum speed that a truck should go on an expressway is

(A) 10 MPH
(B) 40 MPH
(C) 15 MPH
(D) 55 MPH

5. Truck # 8472 dumps a full load of snow and ice at the same dump location at the following times: 8:10 A.M., 9:25 A.M., 10:32 A.M., 11:15 A.M., 12:20 P.M., 1:45 P.M., and 2:30 P.M. The truck dumps 7 cubic yards of snow and ice each time. The total number of cubic yards of snow and ice dumped by this truck is

(A) 35 cubic yards
(B) 42 cubic yards
(C) 49 cubic yards
(D) 70 cubic yards

Answer questions 6 through 8 on the basis of the following information: On a busy day at the landfill, Truck #P-396 dumped full loads of garbage at 7:30 A.M., 8:45 A.M., 9:55 A.M., 11:15 A.M., and 12:25 P.M.; Truck #T-9045 dumped garbage loads at 9:20 A.M., 10:15 A.M., 11:30 A.M., 1:15 P.M., 2:10 P.M., and 3:30 P.M.; Truck #G-3062 dumped at 10:00 A.M., 11:42 A.M., 1:15 P.M., 2:20 P.M., 3:35 P.M., and 4:40 P.M.; Truck #S-556 dumped at 12:00 P.M., 12:56 P.M., 2:25 P.M., 4:05 P.M., and 5:25 P.M.

6. The total number of loads of garbage delivered to the landfill was

(A) 20
(B) 22
(C) 23
(D) 25

7. The truck that dumped the greatest number of loads in the morning was

(A) #P-396
(B) #T-9045
(C) #G-3062
(D) #S-556

8. The truck that did the most work in the afternoon was

(A) #P-396
(B) #T-9045
(C) #G-3062
(D) #S-556

Answer questions 9 and 10 based on the information below.

Sanitation workers follow these rules when plowing snow:

1. Fire hydrants—plow snow for 20 feet on each side of hydrant
2. Bus stops—plow the total length of the stopping areas
3. Steel plates in roadway—lift plow blade
4. Parking meter areas—pile snow on sidewalk, not in front of posts

9. The correct rule for plowing bus stops is to

    (A) pile the snow on the sidewalk
    (B) plow the length of the bus
    (C) plow the total length of the stopping area
    (D) plow 20 feet beyond the length of the bus

10. When plowing an area with steel plates on the street,

    (A) pile the snow on the steel plates
    (B) plow snow 20 feet around the steel plates
    (C) lower the plow to thoroughly clear the plates
    (D) raise the plow blade

Answer questions 11 to 15 based on the following list of sanitation worker uniform items.

Item	Description	Where to Get Them
Work shoes	High top; brown; nonskid	Any store
Rain boots	Black rubber	Any store
Pants	Gray with white stripes	Breen Co.
Shirts	Long sleeves; forest green	Breen Co.
Shirts	Short sleeves; forest green	Breen Co.
Jacket	Gray	Breen Co.
Safety vest	Bright orange with white stripes	Breen Co.
Gloves	Gray canvas	Any store
Cap	Gray heavy duck	Harvey Co.
Raincoat	Yellow with red stripes	Any store
Rainhat	Yellow	Any store
Belt	Black	Any store
Badge	Silver-colored metal	Department

11. Which item on the list can you get from any store?

    (A) long-sleeved shirts
    (B) badge
    (C) safety vest
    (D) belt

12. Which items on the list are yellow?

    (A) pants and caps
    (B) rain boots and work gloves
    (C) raincoats and safety vests
    (D) rainhats and raincoats

13. Which item on the list can you get only from Breen Co.?

    (A) badge
    (B) short-sleeved shirt
    (C) cap
    (D) work gloves

14. Based on the list, which one of the following statements is NOT correct?

(A) You can get work gloves at any store.
(B) You can get your safety vest from the department.
(C) You can get pants from the Breen Co.
(D) You can get your cap at Harvey Co.

15. Which items on the uniform list have white stripes?

(A) pants and safety vests
(B) raincoats and safety vests
(C) long-sleeved shirts and pants
(D) raincoats and rainhats

Answer questions 16 to 24 based on the Daily Work Ticket below.

Daily Work Ticket

Date	Day	Equip. No.	Work Type	Freq.	Location	Route	Shift	1. Gasoline No. of Gals.
7/28	Thurs	G-016	Collection		District 50 Section 501	4	7-3 PM	2. Oil No. of Quarts

Equipment		3. Start of Shift Mileage
Time out _____ Driver _____		4. End of Shift Mileage
Time in _____ Loader _____		5. Total Miles Traveled
Loader _____		

Load No.	Dumping Instructions	Section Supervisor
1st load	Incinerator	Robert Hicks
2nd load	Landfill	Robert Hicks
3rd load	Marine Transfer Station	Robert Hicks
4th load	Return to garage	Robert Hicks

16. On Thursday, July 28, you are assigned as the driver of sanitation collection truck G-016 on the 7–3 shift. Before leaving the garage to cover your assigned route, you check the oil level and find that it is down 4 quarts. You add 4 quarts of oil and fill in the information on your "Daily Work Ticket" in the box numbered

(A) 1
(B) 2
(C) 4
(D) 5

17. As the driver of Equipment No. G-016, how many people will be working with you?

    (A) 1
    (B) 2
    (C) 3
    (D) 4

18. At the beginning of the day, it takes a half hour to pick up your assignment and to check out the condition of the equipment. At the end of the day, checking the equipment back in takes fifteen minutes. In the Equipment Time out and Time in spaces, you should write

    (A) 7:00 A.M. and 3:00 P.M.
    (B) 6:30 A.M. and 3:15 P.M.
    (C) 7:30 A.M. and 2:45 P.M.
    (D) 7:15 A.M. and 2:30 P.M.

19. Before you leave the garage to start your assignment, you read the mileage gage (odometer). The odometer shows 19,902 miles. In which numbered space on the "Daily Work Ticket" should you write this number?

    (A) 1
    (B) 3
    (C) 4
    (D) 5

20. When you return this truck to the garage at the end of your shift, the odometer reads 19,953 miles. How many miles did your truck travel during the assigned shift?

    (A) 51 miles
    (B) 56 miles
    (C) 61 miles
    (D) 69 miles

21. The type of work listed for Equipment G-016 is

    (A) collection
    (B) flushing
    (C) mechanical sweeping
    (D) sanding

22. Sanitation vehicle No. G-016 has been fully loaded three times. The correct dumping instruction is

    (A) return to the garage
    (B) dump load at the landfill
    (C) dump load at the incinerator
    (D) dump load at the Marine Transfer Station

23. The location of Route 4 is

    (A) District 501, Section 50
    (B) District 51, Section 501
    (C) District 50, Section 501
    (D) District 50, Section 510

24. Robert Hicks is

(A) the dumping instructor
(B) a loader
(C) section supervisor
(D) the garage manager

Answer questions 25 to 29 based on the information below.

During snowstorms, sanitation workers spread sand on highways, roads, and city streets. Sand offers extra traction to moving vehicles, though it does not melt the ice and snow. The salt, which is spread after the sand, causes snow and ice to melt. The sand is spread by a special truck known as a "spreader." The sand spreader is the first type of equipment in a snowstorm. A fully loaded sand spreader holds 15 cubic yards of sand. Each cubic yard of sand weighs 1 ton.

25. The first type of equipment placed into service in a snow storm is a

(A) snow plow
(B) salt spreader
(C) collection truck
(D) sand spreader

26. You are a sanitation worker assigned to a sand spreader. One of your duties is to record how many tons of sand are loaded on your truck. When your truck is fully loaded, you must record

(A) 14 tons
(B) 15 tons
(C) 16 tons
(D) 30 tons

27. Which one of the following statements is correct?

(A) Sand spreaders are used to melt crosswalks, bus stops, and fire hydrant areas.
(B) Snow plows are the first equipment put into service to melt snow during snow removal operations.
(C) Salt spreaders are used to melt snow on highways, roads, and city streets.
(D) Snow plow blades must be lowered when crossing railroad tracks or steel plates.

28. The Department of Sanitation spreads sand when

(A) the temperature drops below freezing
(B) snow falls
(C) there has been an accident caused by skidding
(D) snow melts

29. Sanitation workers spread salt to

(A) tow away cars
(B) improve garbage collection
(C) improve traction
(D) melt ice and snow

Answer questions 30 to 35 based on the paragraph below.

At 10:15 A.M. on Wednesday, June 3, a sanitation truck driven by Sanitation Worker LeClair was travelling south on Chapel Street. A car with Connecticut license plate 673 XYZ driven by Pat Johnson was travelling east on Bradford Street. The car ran through a red light and hit the right front bumper of the truck. No one was injured, but the front end of the car was damaged and the truck has a bent right front bumper. At 10:35 A.M. Police Officers Jones and Murphy arrived to investigate.

30. The direction in which the sanitation truck was travelling when the accident occurred is

(A) north
(B) south
(C) east
(D) west

31. The direction in which the car was travelling when the accident occurred is

(A) north
(B) south
(C) east
(D) west

32. What is the location of the accident?

(A) Chapel Street and Bradford Street
(B) Bradford Street and South Street
(C) East Parkway and Chapel Street
(D) South Street and East Parkway

33. The last name of the sanitation driver is

(A) LeClair
(B) Jones
(C) Johnson
(D) Murphy

34. The damage to the truck was to the

(A) left front bumper
(B) right front fender
(C) left front fender
(D) right front bumper

35. The police officers arrived to investigate the accident at

(A) 10:15 A.M.
(B) 10:25 A.M.
(C) 10:35 A.M.
(D) 10:45 A.M.

Answer questions 36 to 42 based on the chart below. This chart shows the location of vehicles and supplies kept in the garage. Sanitation workers consult this chart each day when they receive their assignments. It tells them where to find the vehicles and supplies they will need for work that day.

Location of Vehicles and Supplies

ROW 1-2-3	ROW 4-5	ROW 6
Collection Trucks Sweepers Large Wreckers Dual Purpose Trucks	Pest Control Trucks Front-end Loaders Flushers Small Wreckers	Supervisor Cars Cushman Haulsters Rear Loader Trucks Mechanical Brooms

SECTION A	SECTION B
Long Handle Shovels Push Brooms Can Carriers Short Hoses	Snow Pushers Pan Scrapers Short Handle Shovels Whisk Brooms

36. A sanitation worker assigned to use a small wrecker will find it in

(A) row 6
(B) section B
(C) row 4-5
(D) row 1-2-3

37. A sanitation worker assigned to use a Cushman haulster will find it in

(A) section A
(B) row 1-2-3
(C) section B
(D) row 6

38. A sanitation worker will find a dual-purpose truck in

(A) row 6
(B) row 1-2-3
(C) section B
(D) row 4-5

39. A sanitation worker will find a pan scraper in

(A) section B
(B) section A
(C) row 1-2-3
(D) row 6

40. What items will be found in row 4-5?

(A) short hoses
(B) flushers
(C) rear loader trucks
(D) sweepers

41. Which items will be found in section B?

    (A) whisk brooms and sweepers
    (B) snow pushers and short handle shovels
    (C) Cushman haulsters and dual purpose trucks
    (D) snow pushers and can carriers

42. Which items will be found in row 1-2-3?

    (A) mechanical brooms and sweepers
    (B) rear loader trucks and collection trucks
    (C) large wreckers and collection trucks
    (D) dual purpose trucks and short hoses

Answer questions 43 to 46 based on the following information:

Sanitation trucks have manual transmissions. The transmission is the part of a truck that transmits power to the wheels and makes the wheels turn. The driver of a sanitation truck must move a gear shift handle to the correct position to make sure that the wheels have enough power to move the truck. The driver selects the correct position of the gear shift handle in this way:

Position 1 is used when the truck is moving on a big hill.

Position 2 is used when the truck is moving on a dirt road or is plowing snow.

Position 3 is used when the truck is full and moving on level ground or a small hill.

Position 4 is used when the truck is empty and moving on level ground or a small hill.

43. A driver of a sanitation truck leaves the garage with an empty truck. The route is on level ground and has a few small hills. The correct gear shift handle position to use in this situation is

    (A) position 1
    (B) position 2
    (C) position 3
    (D) position 4

44. A full sanitation truck moving on a small hill should use which gear shift handle position?

    (A) position 1
    (B) position 2
    (C) position 3
    (D) position 4

45. A full sanitation truck plowing snow on a dirt road should use which gear shift handle position?

    (A) position 1
    (B) position 2
    (C) position 3
    (D) position 4

46. An empty sanitation truck moving on a big hill should use which gear shift position?

(A) position 1
(B) position 2
(C) position 3
(D) position 4

47. Sanitation workers wear safety shoes while on duty to protect their feet. When they wear these shoes they have fewer serious injuries.

Based on the statements above, sanitation workers wear safety shoes because they

(A) keep their feet warm
(B) help avoid serious injury to their feet
(C) are more comfortable
(D) are more attractive

Answer questions 48 to 50 based on the following:

A sanitation driver must follow these procedures after a traffic accident.

1. Check to see if anyone involved in the accident is hurt.
2. Exchange driver license information with the driver of the other vehicle.
3. Write down the license plate number of the other vehicle.
4. Ask the driver of the other vehicle to wait for the sanitation garage officer.
5. Do not move the Sanitation Department vehicle unless ordered by a police officer.
6. Do not leave the scene of the accident.
7. Get the names and addresses of people who saw the accident.

48. While you are driving a street sweeper, a car hits the sweeper. You hop out of the sweeper and talk to the driver of the car. Neither of you is injured at all, and there are no passengers in the car. The sweeper is not damaged, and the driver of the car is not concerned about the dent in the front bumper. Under these circumstances, you should

(A) move the sweeper to the side of the road so that it does not block traffic
(B) call the police to arrest the driver who hit you
(C) get on with the work of cleaning the streets
(D) write down the license plate number of the car

49. The sanitation vehicle you are driving is involved in an accident. What is the correct procedure for you to follow?

(A) Take the injured party to a hospital right away.
(B) Move the vehicle so the other driver can check the damage to his or her car.
(C) Call the police.
(D) Move the vehicle when a police officer tells you to.

50. Your sanitation truck is hit by a car. No one is hurt. You have exchanged driver license information with the driver of the car and are waiting for your garage officer to arrive. The next thing you should do is

(A) tell the driver of the car that he is free to leave
(B) get the names and addresses of people who saw the accident
(C) take the vehicle back to the garage
(D) ask the bystanders to clear the area

Answer questions 51 through 54 based on the following information:

Supervisor Cheung gives Sanitation Worker Kline a route sheet. The route sheet lists streets in which the snow must be plowed. The supervisor tells the crew that the streets must be plowed in the order that they are listed. If, however, the crew finds a street blocked by double-parked traffic, it must plow the next street on the route. After the rest of the route is completed, the crew must return to the street that was blocked. At that time, if the street is open to moving traffic, the crew is to plow the street that was previously skipped.

The route sheet given to Sanitation Worker Kline is listed below:

Route Sheet

Number	Street Locations
1	the left side of State Street
2	the left side of Franklin Avenue
3	the left side of 4th Street
4	the right side of Franklin Avenue
5	the right side of 4th Street
6	the right side of State Street
7	both sides of Grant Street
8	both sides of Linden Boulevard

51. The crew has plowed the left side of Franklin Ave. The next street where it must plow is

(A) the left side of 4th Street
(B) both sides of Grant Street
(C) the right side of Franklin Avenue
(D) both sides of Linden Boulevard

52. The snow plow arrives at the right side of Franklin Avenue. It finds that the street is blocked by a moving van. The crew must now

(A) report the blocked street to the supervisor
(B) wait patiently at the corner until the moving van leaves
(C) plow the right side of 4th Street
(D) plow the right side of State Street

53. The plow finishes its assigned route and returns to the right side of Franklin Avenue. This street was blocked before but is now open to traffic. The correct thing for the crew to do is

(A) call Supervisor Cheung for further instructions
(B) plow the left side of 4th Street
(C) issue a summons for traffic violation to the driver of the moving van
(D) plow the right side of Franklin Avenue

54. Number 7 on the route sheet identifies

(A) the right side of State Street
(B) both sides of Linden Boulevard
(C) both sides of Grant Street
(D) the left side of State Street

Answer questions 55 to 60 based on the accident report below.

Sanitation Worker O'Neill injured her wrist while loading garbage. The following is her statement describing how the accident happened: "At 6:40 A.M., I was loading garbage and trash on the north side of Dutch Street between Forrest Avenue and Maple Lane. There were garbage cans filled with rocks, plaster, bottles, and pieces of wood. There was also a pile of rugs. While I was carrying a rug to the truck, I was startled by a loud, sharp car horn blast which caused me to drop the rug and then trip over it. As a result of my fall, my wrist began to hurt and I could not continue my assigned duties. Sanitation Worker Polski, who saw the accident, notified Section Supervisor Green about the accident. Green arrived at the scene at 6:50 A.M."

55. What was Sanitation Worker O'Neill doing at the time of the accident?

(A) driving a truck
(B) carrying a rug
(C) loading cardboard boxes
(D) moving a can

56. As a result of the accident, Sanitation Worker O'Neill hurt her

(A) wrist
(B) back
(C) leg
(D) ankle

57. Where did the accident happen?

(A) at the corner of Forrest Avenue and Maple Lane
(B) on the south side of Maple Lane
(C) on the north side of Dutch Street
(D) on the east side of Dutch Street

58. The accident happened because Sanitation Worker O'Neill

(A) was carrying a rug that was too heavy for her
(B) tripped over a cat
(C) cut herself on a bottle
(D) was startled by a car horn

59. Who saw the accident happen?

(A) Sanitation Supervisor Green
(B) Sanitation Worker Polski
(C) Sanitation Worker Forrest
(D) Sanitation Supervisor O'Neill

60. Sanitation Supervisor Green arrived at the scene of the accident at

(A) 6:30 A.M.
(B) 6:40 A.M.
(C) 6:50 A.M.
(D) 7:00 A.M.

MODEL EXAM II ANSWER KEY

1. A	11. D	21. A	31. C	41. B	51. A
2. D	12. D	22. D	32. A	42. C	52. C
3. C	13. B	23. C	33. A	43. D	53. D
4. B	14. B	24. C	34. D	44. C	54. C
5. C	15. A	25. D	35. C	45. B	55. B
6. B	16. B	26. B	36. C	46. A	56. A
7. A	17. B	27. C	37. D	47. B	57. C
8. D	18. C	28. B	38. B	48. D	58. D
9. C	19. B	29. D	39. A	49. D	59. B
10. D	20. A	30. B	40. B	50. B	60. C

EXPLANATIONS

1. **(A)** The second sentence tells us that the sanitation workers use the route sheet to make a record of the streets which have been cleaned. Then others use that route sheet to make out the Sanitation Department's Street Sweeping Report.

2. **(D)** This is the clear message of the last sentence.

3. **(C)** The speed limit on *all* dirt roads is 10 MPH regardless of whether or not the dirt roads are within city limits.

4. **(B)** See Rule 2.

5. **(C)** Truck #8472 dumped seven full loads each amounting to seven cubic yards. We know there were seven loads because seven different dumping times are given. You can find the total number of cubic yards by multiplying 7 loads by 7 cubic yards per load: $7 \times 7 = 49$ or by adding $7 + 7 + 7 + 7 + 7 + 7 + 7 = 49$.

6. **(B)** Count up all the times that loads were delivered by all four trucks.

7. **(A)** Truck #P-396 dumped four loads in the morning. Truck #T-9045 dumped three loads in the morning; truck #G-3062 dumped only two loads in the morning; and truck #S-556 did all its dumping in the afternoon.

8. **(D)** Truck #S-556 dumped five loads in the afternoon. Truck #G-3062 dumped four loads in the afternoon; truck #T-9045 dumped three loads in the afternoon; and truck #P-396 dumped only one.

9. **(C)** This is rule number 2.

10. **(D)** This is rule number 3.

11. **(D)** Belts can be bought at any store. Shirts and safety vests must be purchased from Breen Co. Badges are issued by the department.

12. **(D)** Raincoats and rainhats are yellow. Pants, caps, and work gloves are gray, and safety vests are bright orange.

13. **(B)** The short-sleeved shirt must be purchased at Breen Co. Gloves can be bought at any store, and caps are available at Harvey Co. Badges are issued by the department.

14. **(B)** The safety vest must be purchased at Breen Co.

15. **(A)** Pants and safety vests have white stripes. Raincoats have red stripes. The other items have no stripes at all.

16. **(B)** You write in "4" for 4 quarts, but you write it into Box 2 which refers to oil.

17. **(B)** 2 loaders work on the collection truck with the driver.

18. **(C)** You check in at 7:00 A.M. then spend a half hour preparing for your route. You then check out your equipment at 7:30 A.M. At the end of the day, knowing that it takes fifteen minutes to wind up and that you can go off duty at 3:00 P.M., you check in your equipment at 2:45 P.M.

Completed Daily Work Ticket

Date	Day	Equip. No.	Work Type	Freq.	Location	Route	Shift	1. Gasoline No. of Gals.
7/28	Thurs	G-016	Collection		District 50 Section 501	4	7–3 PM	2. Oil No. of Quarts 4

Equipment		3. Start of Shift Mileage 19,902
Time out 7:30 A.M. Driver _____		4. End of Shift Mileage 19,953
Time in 2:45 P.M. Loader _____		5. Total Miles Traveled 51
Loader _____		

Load No.	Dumping Instructions	Section Supervisor
1st load	Incinerator	Robert Hicks
2nd load	Landfill	Robert Hicks
3rd load	Marine Transfer Station	Robert Hicks
4th load	Return to garage	Robert Hicks

19. **(B)** Space 3 is the place to record mileage before you set out.

20. **(A)** The difference between 19,953, mileage at the end of the shift, and 19,902, mileage at the start of the shift, is 51 miles. $19,953 - 19,902 = 51$

21. **(A)** This information is given at the top of the work ticket under the heading "Work Type."

22. **(D)** The instructions given on the work ticket are to take the third load to the Marine Transfer Station.

23. **(C)** The information is given at the top of the ticket under "Location." Read the numbers carefully in choosing your answer.

24. **(C)** Again read headings carefully. Robert Hicks is the Section Supervisor.

25. **(D)** The paragraph states: "The Sand spreader is the first type of equipment used at the beginning of a snowstorm." You must answer on the basis of the information in the paragraph even if you disagree.

26. **(B)** The truck holds 15 cubic yards of sand, and each cubic yard weighs one ton, so a fully loaded truck holds 15 tons of sand.

27. **(C)** Salt melts the snow. Sand increases traction. Snow plows neither melt snow nor are they first out in a snowstorm.

28. **(B)** Sand is spread at the start of a snowfall to aid traction.

29. **(D)** Again, you must keep straight the difference between the action of salt and sand. Salt melts ice and snow.

30. **(B)** See the first sentence.

31. **(C)** See the second sentence.

32. **(A)** The sanitation truck was on Chapel Street; the car was on Bradford Street.

33. **(A)** See the first sentence.

34. **(D)** Read carefully. This information is in the next to last sentence.

35. **(C)** See the last sentence.

36. **(C)** The small wreckers are to be found in row 4-5. This series of questions is not difficult. Just read and mark your answer carefully. It is easy to mark the wrong letter when the answer is, for instance, that an item can be found in Section B which is choice (C). Mark the letter of the choice, not the letter of the section.

37. **(D)**

38. **(B)**

39. **(A)**

40. **(B)**

41. **(B)**

42. **(C)**

43. **(D)** This empty truck driving on level or nearly level ground should operate in position 4.

44. **(C)** Since the truck is full, position 3 is needed even on the small hill.

45. **(B)** Full or empty, position 2 is used when a truck is travelling on a dirt road or plowing snow. The fact that this truck is plowing snow on a dirt road is not addressed as a separate situation in the list of rules, so position 2 is correct.

46. **(A)** A big hill requires position 1, even if the truck is empty.

47. **(B)** Safety shoes are worn for the specific purpose of protecting the feet from serious injury. The shoes probably do keep the workers' feet warm and they may even be comfortable, but neither of these is the reason for wearing safety shoes.

48. **(D)** No matter how minor the accident, you must follow the rules. The rules state that you must get the license plate number of the car and wait for your sanitation garage officer.

49. **(D)** The rules tell you not to move the sanitation vehicle unless ordered to do so by a police officer. The rules do not mention what you should do for an injured party, but common sense tells you that an ambulance should be summoned for transportation to a hospital. The rules do tell the sanitation worker not to leave the scene.

50. **(B)** The last in the list of rules states that you should get names and addresses of people who saw the accident. Choice (A) is wrong because Rule 4 states that you should ask the driver of the other vehicle to wait with you for the garage officer.

51. **(A)** Read carefully to keep left and right straight and not skip a line when choosing the answer.

52. **(C)** According to the rules, if a scheduled street is impassable, the crew should skip it for the time being and go directly to the next street on the route. Aftr skipping the right side of Franklin, the next assignment is the right side of 4th Street.

53. **(D)** According to the rules, when the entire route has been completed, the crew should return to any previously blocked streets and plow them if possible.

54. **(C)**

55. **(B)** In the fourth sentence of her statement, O'Neill says that she was carrying a rug.

56. **(A)** In the next sentence she tells that it was her wrist that hurt her after the fall.

57. **(C)** As stated in the first sentence, the collection was being made on the north side of Dutch Street. Forrest Avenue and Maple Lane are mentioned only to help pinpoint the exact location on Dutch Street.

58. **(D)** The rug may have been very heavy, but the accident occurred because O'Neill was startled by the unexpected car horn.

59. **(B)** See the next to last sentence.

60. **(C)** See the last sentence.

ANSWER SHEET FOR MODEL EXAM III

1. Ⓐ Ⓑ Ⓒ Ⓓ   16. Ⓐ Ⓑ Ⓒ Ⓓ   31. Ⓐ Ⓑ Ⓒ Ⓓ   46. Ⓐ Ⓑ Ⓒ Ⓓ

2. Ⓐ Ⓑ Ⓒ Ⓓ   17. Ⓐ Ⓑ Ⓒ Ⓓ   32. Ⓐ Ⓑ Ⓒ Ⓓ   47. Ⓐ Ⓑ Ⓒ Ⓓ

3. Ⓐ Ⓑ Ⓒ Ⓓ   18. Ⓐ Ⓑ Ⓒ Ⓓ   33. Ⓐ Ⓑ Ⓒ Ⓓ   48. Ⓐ Ⓑ Ⓒ Ⓓ

4. Ⓐ Ⓑ Ⓒ Ⓓ   19. Ⓐ Ⓑ Ⓒ Ⓓ   34. Ⓐ Ⓑ Ⓒ Ⓓ   49. Ⓐ Ⓑ Ⓒ Ⓓ

5. Ⓐ Ⓑ Ⓒ Ⓓ   20. Ⓐ Ⓑ Ⓒ Ⓓ   35. Ⓐ Ⓑ Ⓒ Ⓓ   50. Ⓐ Ⓑ Ⓒ Ⓓ

6. Ⓐ Ⓑ Ⓒ Ⓓ   21. Ⓐ Ⓑ Ⓒ Ⓓ   36. Ⓐ Ⓑ Ⓒ Ⓓ   51. Ⓐ Ⓑ Ⓒ Ⓓ

7. Ⓐ Ⓑ Ⓒ Ⓓ   22. Ⓐ Ⓑ Ⓒ Ⓓ   37. Ⓐ Ⓑ Ⓒ Ⓓ   52. Ⓐ Ⓑ Ⓒ Ⓓ

8. Ⓐ Ⓑ Ⓒ Ⓓ   23. Ⓐ Ⓑ Ⓒ Ⓓ   38. Ⓐ Ⓑ Ⓒ Ⓓ   53. Ⓐ Ⓑ Ⓒ Ⓓ

9. Ⓐ Ⓑ Ⓒ Ⓓ   24. Ⓐ Ⓑ Ⓒ Ⓓ   39. Ⓐ Ⓑ Ⓒ Ⓓ   54. Ⓐ Ⓑ Ⓒ Ⓓ

10. Ⓐ Ⓑ Ⓒ Ⓓ   25. Ⓐ Ⓑ Ⓒ Ⓓ   40. Ⓐ Ⓑ Ⓒ Ⓓ   55. Ⓐ Ⓑ Ⓒ Ⓓ

11. Ⓐ Ⓑ Ⓒ Ⓓ   26. Ⓐ Ⓑ Ⓒ Ⓓ   41. Ⓐ Ⓑ Ⓒ Ⓓ   56. Ⓐ Ⓑ Ⓒ Ⓓ

12. Ⓐ Ⓑ Ⓒ Ⓓ   27. Ⓐ Ⓑ Ⓒ Ⓓ   42. Ⓐ Ⓑ Ⓒ Ⓓ   57. Ⓐ Ⓑ Ⓒ Ⓓ

13. Ⓐ Ⓑ Ⓒ Ⓓ   28. Ⓐ Ⓑ Ⓒ Ⓓ   43. Ⓐ Ⓑ Ⓒ Ⓓ   58. Ⓐ Ⓑ Ⓒ Ⓓ

14. Ⓐ Ⓑ Ⓒ Ⓓ   29. Ⓐ Ⓑ Ⓒ Ⓓ   44. Ⓐ Ⓑ Ⓒ Ⓓ   59. Ⓐ Ⓑ Ⓒ Ⓓ

15. Ⓐ Ⓑ Ⓒ Ⓓ   30. Ⓐ Ⓑ Ⓒ Ⓓ   45. Ⓐ Ⓑ Ⓒ Ⓓ   60. Ⓐ Ⓑ Ⓒ Ⓓ

Number Right _____

MODEL EXAM III

60 QUESTIONS—2½ HOURS

Directions: Read each question and choose the best answer. On your answer sheet, darken the letter of the answer you choose.

Answer questions 1 to 3 based on the following paragraph:

In a few years our city will not know what to do with the garbage it collects. Now the garbage is used as landfill, but landfill sites will be all used up within the next eight years. We cannot burn the garbage because burning pollutes the air. We can't dump at sea because of the danger to fish. And shipping our garbage to another state is very expensive. Besides, no other state wants it. Science may have to figure out a new way to dispose of garbage.

1. A good title for this paragraph would be

    (A) "The Difficulty of Getting Rid of Garbage"
    (B) "The High Cost of Getting Rid of Garbage"
    (C) "Different Ways We Can Get Rid of Garbage"
    (D) "Science's Answer to the Garbage Problem"

2. Air pollution results when you

    (A) use garbage as landfill
    (B) ship garbage to another state
    (C) burn garbage
    (D) let garbage lie in the streets

3. When this paragraph says that burning garbage *pollutes* the air, it means that burning garbage

    (A) uses up the air
    (B) dirties the air
    (C) fills the air
    (D) burns up the air

Answer questions 4 to 7 based on the following paragraph:

The secret of good snow removal is being ready before the first flake falls. There must be enough equipment, and the equipment must be in good working order. There must be plenty of sand, and the sand must be accessible—that is, it must be somewhere that the trucks can get to it easily. Each worker must have an assignment beforehand. When it is clear that a snowstorm is about to start, the crew should start sanding even before the storm begins. The sand will keep the roads from becoming too slippery. When the snow begins to accumulate, the plows follow the route of the sanders.

71

4. The paragraph suggests that a sander crew should

    (A) consist of only experienced workers
    (B) be made up of workers who have been specially trained in sanding
    (C) have members who work well together
    (D) be clear as to each worker's assignment

5. The paragraph says that the sand must be accessible. This means that the sand

    (A) must be plentiful and of good quality
    (B) should be in a place where sanders can get to it easily
    (C) should be in good working condition
    (D) should be easy to spread

6. The word *accumulate* as used in the last sentence means

    (A) melt
    (B) mix with the sand
    (C) pile up
    (D) gather

7. We learn from this paragraph that you can sometimes

    (A) start sanding before a storm
    (B) sand and plow at the same time
    (C) learn your assignment when the snow begins to fall
    (D) do a good job without much equipment

Answer questions 8 to 11 based on the following information.

In this state, no person may operate a motor vehicle or motorcycle on public highways without a valid license. Operator licenses of Classes 5 and 6 are reserved for private operators of personal use automobiles and motorcycles. Class 5 and Class 6 license holders cannot drive for any commercial purposes.

    There are four types of chauffeur licenses:

Class 1 authorizes the holder to operate tractor-trailers, truck-trailers, and all other vehicles except motorcycles.

Class 2 authorizes the holder to operate buses and all other vehicles except tractor-trailers, truck-trailers, and motorcycles.

Class 3 authorizes the holder to operate single-unit trucks with a gross weight in excess of 18,000 pounds, taxicabs, and any other vehicle except tractor-trailers, truck-trailers, buses, and motorcycles.

Unclassified authorizes the holder to operate taxicabs and any other vehicles except tractor-trailers, truck-trailers, buses, motorcycles, and single unit trucks with a gross weight in excess of 18,000 pounds.

8. In order to drive a tractor-trailer, the type of chauffeur's license an operator must hold is

    (A) Class 1
    (B) Class 2
    (C) Class 3
    (D) Unclassified

9. An unclassified chauffeur's license will allow its holder to drive

   (A) a bus
   (B) a motorcycle
   (C) a taxicab
   (D) all three

10. If a sanitation truck weighs 21,000 pounds, its driver must have

    (A) a Class 5 operator's license
    (B) an unclassified chauffeur's license
    (C) a Class 1, 2, or 3 chauffeur's license
    (D) passed the motorcycle license examination

11. Which one of the following driver's licenses must a bus driver have?

    (A) unclassified
    (B) Class 2
    (C) Class 3
    (D) none of these

12. The shape of the sign that indicates to a driver that merging traffic has the right of way is

    (A)

    (B)

    (C)

    (D)

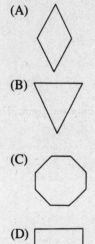

13. This sign means

    (A) no left turn
    (B) right turn allowed after full stop
    (C) no right turn
    (D) left turn only

14. A driver coming upon the following sign on a highway knows that

(A) the road forks to the left
(B) right turns are not permitted
(C) he or she must be alert to traffic merging from the left
(D) he or she must be alert to traffic merging from the right

15. This yellow sign with black letters means that you are coming to a

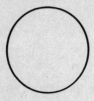

(A) school
(B) railroad crossing
(C) traffic signal
(D) sharp curve

16. One of the following signs has the same meaning to drivers as a flashing red light. That sign is

(A)

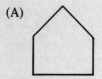

(B)

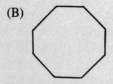

(C)

(D)

17. The double solid lines down the middle of the road as illustrated below mean that a driver

- (A) must not cross the lines except when entering the road from a driveway or leaving the road to go into a driveway
- (B) must not cross the lines under any circumstances
- (C) may cross only to pass
- (D) may cross at any time that it is safe to cross

Questions 18 to 22 refer to the accident report form and the information in the paragraph that follows.

ACCIDENT REPORT

1. Name of person injured in accident _____
2. Date and time of accident _____
3. Nature of injury _____
4. Cause of accident _____
5. Place of accident _____
   Street address _____
   Facility: ( ) garage   ( ) locker room   ( ) road   ( ) office
6. On duty ( )   Off duty ( )
7. Action taken
   ( ) remained on duty   ( ) transported to hospital
   ( ) taken home   ( ) treated by private physician
   ( ) treated by Department physician
8. Name of physician _____
9. Treatment given _____
10. Follow-up advised _____
11. Name of person reporting _____

On Monday, March 18, at 10:45 A.M., while emptying the trash basket at the corner of Jackson Parkway and Green Street, Sanitation Worker Kevin Malone received a deep puncture wound to his upper arm from the broken spoke of an umbrella. The driver of the sanitation vehicle, Majorie Mason, drove Malone back to the Department garage where he was treated by Department Physician Leslie Jonas. Dr. Jonas cleansed the wound and prescribed an antibiotic to be taken for twelve days.

18. The name of the driver, who filled out the form, should appear on line

- (A) 1
- (B) 3
- (C) 8
- (D) 11

19. When completed, line 2 should read

- (A) March 18, 10:45 A.M.
- (B) Monday at 10:45
- (C) March 18th at 10:45
- (D) Monday, March 18, quarter to 11

20. The entry for item 4 reads: "Punctured upper arm on broken umbrella spoke." This entry is

(A) correct
(B) wrong because the wound was to his thigh
(C) wrong because "broken umbrella spoke" should be entered for item 3
(D) wrong because punctured upper arm should be entered for item 3

21. Item 7 should be checked

(A) treated by department physician
(B) treated by private physician
(C) taken home
(D) remained on duty

22. Items 5 and 6 are written up this way:

Jackson Parkway and Green Street
(X) garage
On duty (X)

There is an error in these entries because

(A) item 5 should be checked "road"
(B) item 6 should be checked "off duty"
(C) the trash basket should be mentioned in item 5
(D) the address of the accident should read Jackson Street and Green Parkway

23. City A has 632 workers assigned to snow removal. City B has 468 workers assigned to snow removal. How many more workers are assigned to snow removal in City A?

(A) 164
(B) 176
(C) 276
(D) 1100

24. This year's equipment budget is $57,460. If $46,697 has been spent so far, how much more is left for this year?

(A) $9157
(B) $9763
(C) $9873
(D) $10,763

25. The mileage registered on a sanitation vehicle on April 3rd read 39,354. On December 4th the reading for this same vehicle was 52,453. How many miles did the vehicle travel in that time?

(A) 12,101 miles
(B) 12,199 miles
(C) 13,099 miles
(D) 13,101 miles

26. In the first year of recycling, a town collected 17 tons of newspaper. In its second year, total newspaper tonnage was 25. By the third year, newspaper collected for recycling reached 38½ tons. By how much did the third year's collection exceed that of the second year?

(A) 8 tons
(B) 13½ tons
(C) 21.5 tons
(D) 63 tons

27. A sanitation department spent $42,498.75 last month. Of that amount, $5,324.25 was charged to street cleaning. How much was spent for everything else?

    (A) $37,174.50
    (B) $37,185.00
    (C) $38,075.25
    (D) $38,174.50

28. In the past ten years, the payroll costs in a sanitation district have doubled. If the total payroll ten years ago was $395,536.48, this year it was

    (A) $681,072.96
    (B) $709,073.96
    (C) $791,072.96
    (D) $791,073.96

29. A worker's yearly salary was increased from $19,997.37 to $21,477.36. How much was the worker's raise?

    (A) $1479.99
    (B) $1480.99
    (C) $1580.09
    (D) $2520.01

30. One year the expenses in a district were salaries $424,987; equipment $183,682; repairs $71,401; all other expenses $212,743. What were the total costs to the district that year?

    (A) $792,804
    (B) $803,723
    (C) $892,813
    (D) $902,913

Answer questions 31 to 43 based on the following paragraph:

On Friday, February 2, at 8:30 A.M. George White began to plow the right eastbound lane of the Highland Avenue Bridge. It was snowing heavily, and the surface of the roadway was slippery. At 8:32 A.M. White saw a westbound station wagon skid and strike a westbound sedan. Both cars were badly damaged. The station wagon overturned and came to rest 8 feet from the median. The driver of the station wagon, Hortense Correa, was thrown clear and landed in the middle of the road. The other car smashed against the barrier. The driver of the sedan, Marcus Weinberg, was pinned behind the steering wheel and suffered cuts about the face. The bridge operator, Frank Smith, telephoned for an ambulance. First aid was given to both drivers. They were taken to Mercy Hospital by an ambulance that arrived on the scene at 9:07 A.M. driven by Kenji Yamata. Police Officer Robin Wint, Badge No. 71162, had arrived before the ambulance and recorded all the details of the accident including the statements of George White and of Charlene Kozlowski, another eyewitness.

31. The time of the accident was

    (A) 8:30 A.M.
    (B) 8:32 A.M.
    (C) 8:32 P.M.
    (D) 9:07 A.M.

32. The plow was being operated by

(A) Frank Smith
(B) Frank White
(C) George Smith
(D) George White

33. The accident involved

(A) a sedan and a station wagon
(B) a sedan, a station wagon, and the snowplow
(C) a sedan, a station wagon, and an ambulance
(D) a sedan and two station wagons

34. The person named Charlene Kozlowski was

(A) a police officer
(B) an eyewitness
(C) an ambulance driver
(D) driver of the station wagon

35. The time that elapsed between the accident and the arrival of the ambulance was

(A) 7 minutes
(B) 28 minutes
(C) 32 minutes
(D) 35 minutes

36. The weather was

(A) foggy
(B) slippery
(C) sleety
(D) snowy

37. The station wagon was being driven by

(A) Hortense Weinberg
(B) Marcus Correa
(C) Charlene Correa
(D) Hortense Correa

38. Marcus Weinberg was the

(A) driver of the ambulance
(B) driver of the sedan
(C) bridge operator
(D) police officer

39. The barrier was

(A) struck by the sedan
(B) struck by the station wagon
(C) struck by both cars
(D) struck by neither car

40. The damage to

    (A) the snowplow was slight
    (B) the sedan was severe, but that done to the station wagon was slight
    (C) the station wagon was severe, but that done to the sedan was slight
    (D) both cars were severe

41. The woman driver, Hortense Correa,

    (A) was pinned behind the wheel
    (B) suffered face cuts
    (C) was not wearing a seat belt
    (D) was taken to County Hospital

42. The name of the bridge operator was

    (A) Frank Smith
    (B) Robin Smith
    (C) Frank Wint
    (D) Robin Wint

43. When Kenji Yamata arrived at the scene of the accident

    (A) he called for an ambulance
    (B) Robin Wint was already there
    (C) an ambulance was already there
    (D) Marcus Weinberg had left

44. The word on this sign means that the driver should

    (A) stop
    (B) turn around
    (C) let merging traffic enter the roadway
    (D) look carefully before proceeding into the flow of traffic

45. What should you look for when you see this sign?

    (A) hitchhikers
    (B) schoolchildren
    (C) a parade
    (D) a flagger

46. If you see a sign of this shape, you will have to

(A) stop
(B) take a detour
(C) slow down
(D) turn off your two-way radio

47. A sign of this shape gives

(A) instructions which must be followed
(B) useful information about the road ahead
(C) safety recommendations
(D) information about routes and destinations

48. A sanitation worker will most likely create the best "image" of the Department of Sanitation in the mind of the public by

(A) answering all questions
(B) avoiding arguments
(C) doing the job well
(D) minding his or her own business

49. The best way to move this object 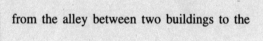 from the alley between two buildings to the street is

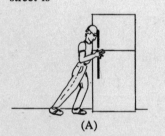
(A)

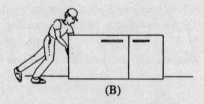
(B)

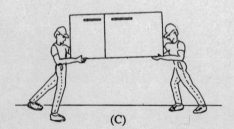
(C)

(D)

50. A sign of this shape

(A) has the force of law
(B) serves as a warning
(C) does not exist
(D) points out tourist attractions

51. In an accident report the *least* useful of the following information would be the

(A) name of the injured worker
(B) time and date accident occurred
(C) length of time the injured worker has been employed by the sanitation department
(D) extent of the injury

52. If most accidents occur in the last hour of the workday, they are most likely due to

(A) carelessness on the part of the workers
(B) improper training
(C) lack of interest in safety on the part of the workers
(D) mental and physical fatigue

53. If you observe what appears to be a serious accident to a coworker, your first course of action should be to

(A) seek out the supervisor
(B) fill out an accident report
(C) notify the injured worker's family
(D) render any possible first aid

54. A report of an accident should be made as soon after the occurrence as practical chiefly because

(A) the witnesses may change their stories before the report is written
(B) the details of the accident will still be fresh in the reporter's mind
(C) the reporter should get the report in so that he or she can get back to work
(D) speed is more important than completeness or accuracy

55. In order to avoid injuries, it is proper to lift heavy objects

(A) with the knees bent and the back held stiff
(B) with both the knees and the back held stiff
(C) using the arms only
(D) in the manner best suited to the individual doing the lifting

56. If a worker who is to perform a job in a hurry notices that a piece of equipment required for the job is defective, the worker should

(A) hold off completing the job until the equipment is either replaced or repaired.
(B) go ahead with the job because it must be completed in a hurry
(C) consult with a coworker
(D) be cautious while using the defective equipment

57. If after completing an accident report you discover an additional piece of information which may influence the conclusions drawn from the report, you should

(A) ignore the information because the report has already been completed and it must be submitted promptly
(B) rewrite the report including the additional information you have discovered
(C) submit the original report and wait to see if the additional information is really important
(D) ask the injured worker for an opinion on the usefulness of the information

58. As a sanitation worker you notice a piece of equipment which you believe can help you with your work. The equipment is somewhat complicated, and you have never received any instruction as to its operation, although other workers have. You should

(A) use the equipment because you will do a better job
(B) put off doing your work until you can receive instruction
(C) ask another worker to teach you to use the equipment
(D) do your work without using the equipment until you have been taught how to use it

59. Departmental regulations require you to wear heavy leather protective work gloves when working on a collection truck, but you have found that the gloves make your hands hot and uncomfortable. The best thing for you to do is

(A) "forget" the gloves at home
(B) tell your supervisor that you would prefer not to wear the gloves
(C) wear the gloves
(D) carry the gloves and slip them on when you think someone is looking

60. You are driving the salt spreader during a heavy snowstorm and find that your vehicle is skidding on the roadway. You should

(A) speed up and get the job done as quickly as possible before road conditions worsen further
(B) return to the garage at once; it is unsafe to be driving
(C) slow down and drive more cautiously
(D) pull over and wait for the worst of the storm to pass

MODEL EXAM III ANSWER KEY

1. A	11. B	21. A	31. B	41. C	51. C
2. C	12. B	22. A	32. D	42. A	52. D
3. B	13. C	23. A	33. A	43. B	53. D
4. D	14. A	24. D	34. B	44. D	54. B
5. B	15. B	25. C	35. D	45. D	55. A
6. C	16. B	26. B	36. D	46. C	56. A
7. A	17. A	27. A	37. D	47. A	57. B
8. A	18. D	28. C	38. B	48. C	58. D
9. C	19. A	29. A	39. A	49. D	59. C
10. C	20. D	30. C	40. D	50. B	60. C

EXPLANATIONS

1. **(A)** The paragraph is all about the problems of getting rid of garbage. It mentions a number of possible ways but then gives reasons why each way will not work. It suggests that science may have to find the answer, but it does not say that science has found the solution yet.

2. **(C)** See the third sentence.

3. **(B)** The air pollution created by burning garbage includes ashes and soot and various chemicals.

4. **(D)** The paragraph says, in the fourth sentence, that each worker should have an assignment beforehand. This means that every member of the crew should know what he or she is to do and that each member should know what to expect of the other members.

5. **(B)** The third sentence defines the word *accessible*.

6. **(C)** *Gather* is a dictionary definition of *accumulate*, but the last sentence refers specifically to accumulation of snow. When snow accumulates it piles up.

7. **(A)** The fifth sentence says that the crew should start sanding even before the storm begins.

8. **(A)** "Class 1 authorizes the holder to operate tractor-trailers, truck-trailers, and all other vehicles except motorcycles."

9. **(C)** "Unclassified authorizes the holder to operate taxicabs and any other vehicles except tractor-trailers, truck-trailers, buses, motorcycles, and single unit trucks with a gross weight in excess of 18,000 pounds."

10. **(C)** A class 3 license authorizes the holder to operate single-unit trucks with a gross weight in excess of 18,000 pounds. A sanitation truck is a single-unit truck.

11. **(B)** The bus driver needs a class 2 license.

12. **(B)** This is the shape of the YIELD sign. The yield sign indicates to a driver that he or she should allow traffic that is entering the roadway to have the right of way.

13. **(C)** A slash through a symbol means that the activity indicated by the symbol is prohibited. The symbol represents a right-turn arrow. This sign means that right turns are not allowed.

14. **(A)** A sign containing a symbol is read from the bottom up. A driver seeing this sign can expect a road to lead to the left from the major roadway. If the sign were to indicate that a road was merging from the left, it would look like this:

; and if a road were merging from the right, like this:

15. **(B)**

Color: Yellow with black letters "RR" and "X" symbol.

Meaning: There is a railroad crossing ahead. You should use caution, and you may have to stop.

16. **(B)** This is the shape of a STOP sign. Both a flashing red light and a stop sign indicate that the driver must come to a full stop, look both ways, then proceed when safe.

17. **(A)** Double solid lines down the middle of the road, either yellow or white, serve to restrict crossing of the road. On a two-lane highway, the double lines mean "No Passing." Double lines may, however, be crossed by vehicles leaving the roadway to enter a driveway on the far side of the road and by vehicles entering the roadway from a driveway and turning left.

18. **(D)** The driver of the sanitation vehicle, Marjorie Mason, filled out the form. Therefore, Marjorie Mason was the person reporting and her name should appear on line 11.

19. **(A)** This entry gives both date and precise time. It is important that the report indicate whether the accident occurred in the morning or at night.

20. **(D)** Line 4 asks only for the cause of the injury. The nature of the injury belongs on line 3.

21. **(A)** Leslie Jonas was the department physician who treated Malone. We have no information as to whether Malone then remained on duty or went home.

22. **(A)** The incident occurred at the corner of Jackson Parkway and Green Street and not in the garage (nor in the locker room or in an office). The item is best checked "road."

The completed accident report should look like this:

ACCIDENT REPORT

1. Name of person injured in accident Kevin Malone
2. Date and time of accident March 18, 10:45 A.M.
3. Nature of injury puncture wound to upper arm
4. Cause of accident broken umbrella spoke
5. Place of occurrence
   Street address Jackson Parkway at Green Street
   Facility ( ) garage ( ) locker room (X) road ( ) office
6. On duty (X) Off duty ( )
7. Action taken
   ( ) remained on duty   ( ) transported to hospital
   ( ) taken home   ( ) treated by private physician
   (X) treated by Department physician
8. Name of physician Leslie Jonas
9. Treatment given wound cleansed—antibiotics prescribed
10. Follow-up advised take antibiotic for twelve days
11. Name of person reporting Marjorie Mason, driver

23. **(A)** 632   workers in city A
       $-\underline{468}$ workers in City B
        164   more in City A

24. **(D)** $57,460   budgeted
       $-\underline{46,697}$ spent
         $10,763   left

25. **(C)** 52,453   December 4 reading
       $-\underline{39,354}$ April 3 reading
        13,099   travelled between April 3 and December 4

26. **(B)** 38½ or 38.5 tons collected in third year
        $-\underline{25}$ tons collected in second year
        13.5 or 13½ more tons collected in third year than in second year

Note that the information given about newspaper collection in the first year has nothing to do with the question. Ignore it.

27. **(A)** $42,498.75   total spent
       $-\underline{5,324.25}$ charged to street cleaning
        $37,174.50   spent for everything other than street cleaning

28. **(C)** $395,536.48 × 2 = $791,072.96

        or $395,536.48
        $+\underline{395,536.48}$
          $791,072.96

29. **(A)**

$21,477.36	new salary
− 19,997.37	old salary
$ 1,479.99	raise

30. **(C)**

$424,987	salaries
183,682	equipment
71,401	repairs
+ 212,743	other expenses
$892,813	total costs

31. **(B)** The correct answer is found in the third sentence. 8:30 A.M. is the time George White began to plow; 9:07 A.M. is the time the ambulance arrived.

32. **(D)** See the first sentence. Read carefully to keep names straight.

33. **(A)** See the third sentence.

34. **(B)** See the last sentence.

35. **(D)** The accident occurred at 8:32 A.M., and the ambulance arrived at 9:07 A.M.

$$9:07 = 8:67$$
$$- 8:32 = -8:32$$
$$35 \text{ minutes}$$

36. **(D)** The second sentence tells us that it was snowing heavily. There may have been fog and sleet as well, but we have no way of knowing this from the paragraph. The roadway was slippery; weather cannot be slippery.

37. **(D)** Read carefully. Look back at the paragraph as needed.

38. **(B)** Reread the paragraph to be sure.

39. **(A)** The station wagon overturned and came to rest 8 feet from the barrier. The other car, namely the sedan, was smashed against the barrier.

40. **(D)** The fourth sentence states, "Both cars were badly damaged."

41. **(C)** Hortense Correa was thrown clear of her station wagon and landed in the middle of the road. She would not have been thrown out of her vehicle if she had been wearing a seat belt.

42. **(A)** The bridge operator, Frank Smith, called for an ambulance. Robin Wint was the police officer who reported to the scene.

43. **(B)** Kenji Yamata was the ambulance driver. The police officer, Robin Wint, had arrived on the scene before the ambulance appeared.

44. **(D)** The basic meaning of the YIELD sign is "let the other driver go first." This sign is most often seen at the head of an entrance ramp to a highway. It tells the driver who is about to enter the highway to look carefully and to enter only when safe. Traffic already moving along the roadway has the right of way. Choice (C) has the meaning backwards.

45. **(D)** The sign warns that there is construction ahead and you should be alert for special instructions from a flagger.

46. **(C)** This sign is a fluorescent or reflective orange triangle. It is atached to slow-moving vehicles like tractors or horse-drawn buggies. If you are behind it, you will have to slow down and watch for a safe moment to pass.

47. **(A)** The vertical rectangular sign gives commands like:
"Do Not Pass," "No Turn on Red," "Speed Limit 50," "Keep Right of Divider," and "No Parking." By law you must obey such signs.

48. **(C)** The best public relations are created by the job well done.

49. **(D)** A heavy, bulky object is best moved on a dolly.

50. **(B)** Diamond-shaped signs give useful information about what to expect on the road ahead. They serve as warnings. The symbols on these signs give messages such as:

DIVIDED HIGHWAY ENDS

HILL AHEAD

RIGHT LANE ENDS
MERGE LEFT

TRAFFIC SIGNAL
AHEAD

TWO-WAY TRAFFIC

51. **(C)** This is a common sense question. Obviously, the name of the injured worker, the time and date of the accident, and the nature of the injury are more important.

52. **(D)** You are most likely to have an accident when you are tired. The other choices have little to do with the time of day.

53. **(D)** The key word here is "serious." If the accident is serious, the injured person is in trouble and needs help. First aid comes first. Then get a supervisor.

54. **(B)** Details fade quickly from memory. The sooner the accident report is made the more accurate and complete it will be.

55. **(A)** This method distributes the stress equally between the back and the legs.

56. **(A)** The use of defective equipment is the cause of many accidents. Even caution will not always prevent accidents if defective equipment is being used. Do not use defective equipment.

57. **(B)** An incomplete report can do more harm than good. Take the time to be complete and accurate so as to help prevent future accidents.

58. **(D)** Never use any equipment until you have been trained in its operation by a person qualified and authorized to train you. However, you have a job to do, so do it without the additional equipment. When you have completed your work, you might speak to your supervisor and request training in use of the equipment.

59. **(C)** Safety rules are for your protection. Wear the gloves even if they make you uncomfortable.

60. **(C)** Your job is to salt the roads to make them safe for others. You cannot quit now. But you must also consider your own safety. Slow down and drive very carefully.

ANSWER SHEET FOR MODEL EXAM IV

1. Ⓐ Ⓑ Ⓒ Ⓓ   16. Ⓐ Ⓑ Ⓒ Ⓓ   31. Ⓐ Ⓑ Ⓒ Ⓓ   46. Ⓐ Ⓑ Ⓒ Ⓓ

2. Ⓐ Ⓑ Ⓒ Ⓓ   17. Ⓐ Ⓑ Ⓒ Ⓓ   32. Ⓐ Ⓑ Ⓒ Ⓓ   47. Ⓐ Ⓑ Ⓒ Ⓓ

3. Ⓐ Ⓑ Ⓒ Ⓓ   18. Ⓐ Ⓑ Ⓒ Ⓓ   33. Ⓐ Ⓑ Ⓒ Ⓓ   48. Ⓐ Ⓑ Ⓒ Ⓓ

4. Ⓐ Ⓑ Ⓒ Ⓓ   19. Ⓐ Ⓑ Ⓒ Ⓓ   34. Ⓐ Ⓑ Ⓒ Ⓓ   49. Ⓐ Ⓑ Ⓒ Ⓓ

5. Ⓐ Ⓑ Ⓒ Ⓓ   20. Ⓐ Ⓑ Ⓒ Ⓓ   35. Ⓐ Ⓑ Ⓒ Ⓓ   50. Ⓐ Ⓑ Ⓒ Ⓓ

6. Ⓐ Ⓑ Ⓒ Ⓓ   21. Ⓐ Ⓑ Ⓒ Ⓓ   36. Ⓐ Ⓑ Ⓒ Ⓓ   51. Ⓐ Ⓑ Ⓒ Ⓓ

7. Ⓐ Ⓑ Ⓒ Ⓓ   22. Ⓐ Ⓑ Ⓒ Ⓓ   37. Ⓐ Ⓑ Ⓒ Ⓓ   52. Ⓐ Ⓑ Ⓒ Ⓓ

8. Ⓐ Ⓑ Ⓒ Ⓓ   23. Ⓐ Ⓑ Ⓒ Ⓓ   38. Ⓐ Ⓑ Ⓒ Ⓓ   53. Ⓐ Ⓑ Ⓒ Ⓓ

9. Ⓐ Ⓑ Ⓒ Ⓓ   24. Ⓐ Ⓑ Ⓒ Ⓓ   39. Ⓐ Ⓑ Ⓒ Ⓓ   54. Ⓐ Ⓑ Ⓒ Ⓓ

10. Ⓐ Ⓑ Ⓒ Ⓓ   25. Ⓐ Ⓑ Ⓒ Ⓓ   40. Ⓐ Ⓑ Ⓒ Ⓓ   55. Ⓐ Ⓑ Ⓒ Ⓓ

11. Ⓐ Ⓑ Ⓒ Ⓓ   26. Ⓐ Ⓑ Ⓒ Ⓓ   41. Ⓐ Ⓑ Ⓒ Ⓓ   56. Ⓐ Ⓑ Ⓒ Ⓓ

12. Ⓐ Ⓑ Ⓒ Ⓓ   27. Ⓐ Ⓑ Ⓒ Ⓓ   42. Ⓐ Ⓑ Ⓒ Ⓓ   57. Ⓐ Ⓑ Ⓒ Ⓓ

13. Ⓐ Ⓑ Ⓒ Ⓓ   28. Ⓐ Ⓑ Ⓒ Ⓓ   43. Ⓐ Ⓑ Ⓒ Ⓓ   58. Ⓐ Ⓑ Ⓒ Ⓓ

14. Ⓐ Ⓑ Ⓒ Ⓓ   29. Ⓐ Ⓑ Ⓒ Ⓓ   44. Ⓐ Ⓑ Ⓒ Ⓓ   59. Ⓐ Ⓑ Ⓒ Ⓓ

15. Ⓐ Ⓑ Ⓒ Ⓓ   30. Ⓐ Ⓑ Ⓒ Ⓓ   45. Ⓐ Ⓑ Ⓒ Ⓓ   60. Ⓐ Ⓑ Ⓒ Ⓓ

Number Right _____

MODEL EXAM IV

60 QUESTIONS—2½ HOURS

Directions: Read each question and choose the best answer. On your answer sheet, darken the letter of the answer you choose.

1. When the yellow light goes on in a three-colored traffic light at an intersection, it means that a driver approaching the intersection

   (A) should realize that the red light will come on in seconds
   (B) must be cautious going through the intersection
   (C) may enter the intersection only if making a right-hand turn
   (D) can expect the light to turn green very soon

2. A diamond-shaped highway sign which shows an arrow curving sharply to the right is a warning that

   (A) there is a sharp narrowing of the road ahead
   (B) there is a sharp turn ahead in the road
   (C) right turns are prohibited
   (D) traffic will be merging from the right

3. An injured person should be kept lying down. If the person begins to vomit, his or her head should be turned to one side. The main reason for turning the head to one side is to

   (A) control the spread of germs
   (B) reduce the area of soiled clothing
   (C) keep the person from having a fit
   (D) help prevent choking

4. In general, the problem of snow removal in a big city is much more difficult than it is out in the country because

   (A) snow becomes very dirty in the city
   (B) more snow falls in the city
   (C) city traffic is heavy
   (D) there is greater area to be cleared in the country

5. On occasion the Sanitation Department uses chemically treated river water for flushing streets. The main reason for treating this water before using it for flushing is to

   (A) prevent it from freezing
   (B) kill germs
   (C) make the water more soluble
   (D) increase the cost of flushing

Answer questions 6 to 10 based on the table below.

Location of Frequently Used Sanitation Equipment

District	Collection Trucks	Sweepers	Flushers	Snow Plows	Spreaders
1	480	72	64	312	146
2	125	22	19	150	83
3	376	56	44	280	102
4	196	30	19	170	77

6. According to the above table, there are two districts that have the same number of

    (A) snow plows
    (B) collection trucks
    (C) flushers
    (D) spreaders

7. According to the table, District 2 does *not* have the smallest number of

    (A) collection trucks
    (B) sweepers
    (C) snow plows
    (D) spreaders

8. The total number of sweepers owned by this city is

    (A) 180
    (B) 178
    (C) 182
    (D) 190

9. If District 3 were to acquire as many snowplows as are currently kept in District 1, it would need to get how many more?

    (A) 592
    (B) 312
    (C) 192
    (D) 32

10. District 1 has the greatest population; District 3 is the next in population; then comes District 4; and District 2 has the smallest population. It is reasonable that the amount of equipment in a district should be related to the size of its population. In which districts is the distribution of equipment out of line?

    (A) District 1 and District 2
    (B) District 2 and District 4
    (C) District 1 and District 3
    (D) District 3 and District 4

Answer questions 11 to 13 based on the pictures below.

(1)   (2)   (3)

11. Driver (1) is

   (A) checking to see if it is raining
   (B) signalling the car behind to stop
   (C) giving directions to a pedestrian
   (D) signalling that he wishes to make a left turn

12. Driver (2) is

   (A) signalling that he intends to turn right
   (B) resting his tired arm
   (C) signalling the car behind to stop
   (D) signalling the car behind to pass

13. Driver (3) is

   (A) waving to a friend
   (B) signalling that he wishes to make a right turn
   (C) signalling the driver behind to pass
   (D) getting a suntan

14. Carbon monoxide

   (A) is colorless
   (B) has a very strong odor
   (C) is harmless
   (D) is never found in sewers

15. When calling a physician to give aid to an injured person, the most important information
    to give would be the

   (A) location of the injured person
   (B) kind of first aid being given
   (C) nature and cause of the injury
   (D) supplies available at the scene of the accident

16. The reason to flush a street with a hose downhill rather than uphill is to

   (A) spread the dirt over a larger area
   (B) keep the hose off the street
   (C) permit the water to go over the same place twice
   (D) allow the dirt to be carried away

Answer questions 17 to 21 based on the following paragraph taken from a safety report.

"Some workers are hit by cans thrown by others. The answer is better 'team play' on the part of the workers. Many injuries are due to handling rough or sharp objects. There are more of these cases in the warmer weather when the workers are in the habit of working without gloves, and the remedy lies in instructing them to protect their hands with hand leathers or leather gloves at all times. Accidents due to dropping cans, sewer covers, and slice bars on their toes cause many fractures and severe bruises. To guard against this, the workers have been instructed to wear safety shoes with steel toe-cap inserts. Good shoes also help protect the feet against nails, glass, and other sharp objects in the rubbish through which they must walk. During the past winter, our workers wore new-style caps with earlaps attached, and there were only four cases of frozen ears despite the severity of the weather."

17. "Team play" is suggested as the way to avoid

    (A) frozen ears
    (B) being hit by cans
    (C) broken toes
    (D) cut hands

18. Injuries to the hands are most likely to occur

    (A) when sewer covers are dropped
    (B) in severe weather
    (C) in summer
    (D) when workers wear safety shoes

19. A type of injury *not* mentioned in the paragraph is

    (A) frozen ears
    (B) bruised toes
    (C) strained backs
    (D) cut feet

20. A result of workers' wearing the proper protective clothing will be

    (A) better team play
    (B) fewer injuries
    (C) fewer accidents
    (D) more dropping of cans

21. The feature of safety shoes that protects from dropped sewer covers is

    (A) thick, skid-proof soles
    (B) heavy leather
    (C) high tops
    (D) steel toe-cap inserts

22. When snow falls, the Sanitation Department makes every effort to remove it as quickly as possible. The main reason for speed is to

    (A) prevent the snow from becoming an eyesore
    (B) free the streets for traffic
    (C) demonstrate the department's efficiency
    (D) make use of the mechanical equipment for which the taxpayers have paid

23. Sometimes the Sanitation Department flushes streets with treated river water instead of reservoir water. The chief reason for doing this is to

(A) conserve drinking water
(B) prevent water from freezing on the streets
(C) wash snow off the streets
(D) make it easier to sweep the streets

24. Of the following, the most important reason for clearing waste material from city streets is to

(A) improve the appearance of the city
(B) furnish employment for a large number of employees
(C) improve traffic
(D) remove possible breeding places for rodents

25. You are flushing a street and accidentally splash water on a passing pedestrian who glares at you. The best procedure is to

(A) ignore the incident
(B) tell the pedestrian he should have been looking where he was going
(C) tell the pedestrian that it was not your fault
(D) apologize

26. A flashing yellow light means

(A) stop
(B) the light is about to turn red
(C) caution
(D) the light is about to turn green

27. This sign means

(A) crosswalk
(B) school crossing
(C) you are in the vicinity of a nursing home
(D) children at play

28. If you are making a collection in the vicinity of this sign, you should

(A) blow your horn to announce your arrival
(B) try not to bang the cans
(C) make a U-turn
(D) watch for the detour that is coming up

29. Which of the signs below designates an interstate highway?

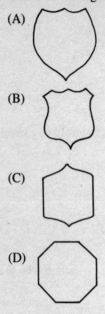

(A)

(B)

(C)

(D)

Answer questions 30 to 35 based on the following passage:

In some ways the job of sanitation worker has grown a lot tougher over the years, primarily because larger trucks with compactors make for far heavier loads for the workers. Instead of the 2½ tons in the old trucks, the new vehicles hold up to 12 tons. As a result, the routes are more than twice as long.

On the other hand, there have also been improvements in the job. Forty years ago most homes were heated with coal, and there would be heavy cans of coal ash to lift and empty into the trucks. Aside from the weight of the ashes, fires inside garbage trucks were a frequent problem. Fire in the truck would mean driving to a firehouse to have water poured on the smoldering pile, often having to first empty the trucks's contents onto the ground and then having to reload it by shovel.

30. Compactors on garbage trucks have made the job harder for sanitation workers because

(A) loads must be put on by shovel
(B) they create a fire hazard
(C) the loads are heavier
(D) of the buildup of ashes

31. New garbage trucks can carry up to _____ tons more garbage than could the old trucks.

(A) 9½
(B) 10½
(C) 12
(D) 14½

32. Not only do the new trucks carry heavier loads, they are also

(A) cleaner
(B) larger
(C) more durable
(D) easier to drive

33. The way that coal ash was gotten into the trucks was by

    (A) pouring water on it
    (B) dumping it onto the ground and shoveling it
    (C) compacting
    (D) lifting the cans and dumping them

34. Because the new trucks hold much more garbage than the old ones

    (A) sanitation workers do not have to work as hard
    (B) each sanitation worker must work a longer day
    (C) sanitation routes have grown longer
    (D) there are no longer fires in the trucks

35. The passage states that fires inside garbage trucks were a frequent problem. As used in this sentence, *frequent* means

    (A) serious
    (B) common
    (C) amusing
    (D) time-consuming

36. Sanitation Worker Daniels, driving the street-cleaning machine, finds an old, cream-colored two-door Mercury with a tattered roof and no license plates along the side of Grand Street between Martin Place and Orange Street. The same car was in the same position three weeks ago, on August 14th, when Daniels last cleaned the same street. Which of the following statements gives this information to the supervisor most clearly and accurately?

    (A) An abandoned car is keeping me from cleaning Grand Street. It is a wreck of a cream-colored Mercury.
    (B) There is a beat-up car at Martin Place and Grand Street and Orange Street, and I haven't been able to clean it for three weeks.
    (C) Someone took the license plates from a yellow car on Grand Street between Martin Place and Orange Streets on August 14th. It's still there.
    (D) An old, battered Mercury with no plates appears to have been abandoned on Grand Street between Martin Place and Orange Street. It has been there since at least August 14th when I last tried to clean the street.

37. Department rules are very specific about what must be done in certain situations and the order in which actions must be taken. In case of collision between a sanitation vehicle and another vehicle, the driver must:

    1. Check to make certain that no occupant of the other vehicle has been injured.
    2. Pull the sanitation vehicle out of the flow of traffic, if possible.
    3. Summon a police officer.
    4. Exchange driver, registration, and insurance information with the driver of the other vehicle.

In a narrow street lined with parked cars on both sides, the driver in front of a collection truck being driven by Phil Cabrini stops short, and Cabrini's truck rear-ends the car. Cabrini jumps out of the truck and goes to the car. He sees that neither the driver nor the passenger has been injured. Cabrini should now

    (A) summon a police officer
    (B) pull the collection truck out of the flow of traffic
    (C) yell at the driver for not signalling
    (D) exchange driver, registration, and insurance information with the car's driver.

Answer questions 38 to 40 based on the map below. The flow of traffic is indicated by the arrows, you must follow the flow of the traffic.

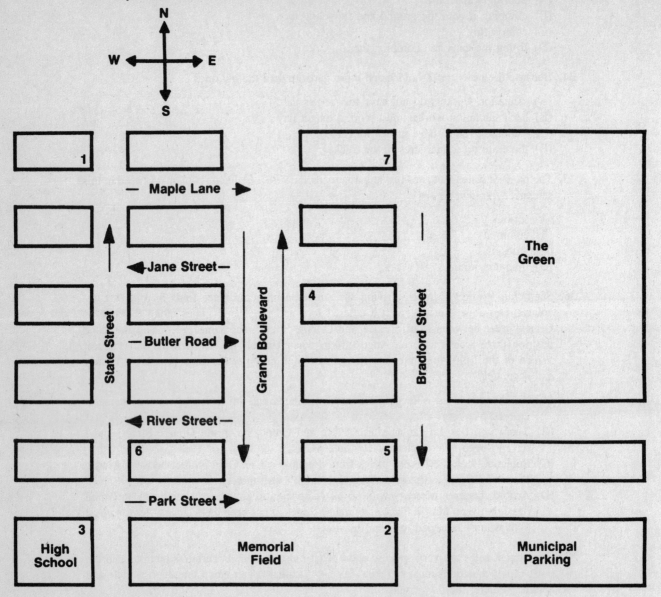

38. If you are located at point (1) and travel east one block, then turn right and travel for four blocks, and then turn left and travel one block, you will be closest to point

(A) 6
(B) 5
(C) 3
(D) 2

39. You have just emptied a wire trash basket at the corner of Butler Road and Bradford Street and must now empty a basket at the corner of Butler Road and Grand Boulevard. Which is the best route to take to the next basket?

(A) Go one block south on Bradford Street, then go one block west on River Street, then go one block north on Grand Boulevard.

(B) Go one block south on Bradford Street, then go two blocks west on River Street, then go one block north on State Street, and then go one block east on Butler Road.

(C) Go one block north on Bradford Street, then go one block west on Jane Street, and then go one block south on Grand Boulevard.

(D) Go two blocks south on Bradford Street, then go one block west on Park Street, then go two blocks north on Grand Boulevard.

40. You have completed your assigned route for spreading salt and still have half a load on your truck. You are at the high school when you receive word that the spreader working Bradford Street is along The Green at the corner of Maple Lane and has run out of salt. What is the best route for you to take to get to that spreader's location to complete its assignment?

   (A) Go two blocks east on Park Street, then go four blocks north on Bradford Street.
   (B) Go two blocks north on State Street, then go two blocks east on Butler Road, then go two blocks north on Bradford Street.
   (C) Go four blocks north on State Street, then go two blocks east on Maple Lane.
   (D) Go one block east on Park Street, then go four blocks north on Grand Boulevard, then go one block east on Maple Lane.

41. "Because of recently adopted changes in the state's requirements for commercial driver's licenses, the Personnel and Sanitation Departments are worried that they may be caught short if too few drivers are able to obtain the new licenses which are necessary to drive sanitation trucks." This quotation means that

   (A) a person who drives a sanitation truck must be a good driver
   (B) you must hold a commercial driver's license in order to drive a sanitation truck
   (C) very few drivers of sanitation trucks can qualify for a commercial driver's license
   (D) the Sanitation Department is short of drivers

42. A commercial vehicle is a vehicle that weighs more than 26,000 pounds when loaded or a lighter vehicle that carries hazardous materials. The driver of a commercial vehicle must have a commercial vehicle driver's license. For which of the following vehicles does the driver *not* need a commercial vehicle driver's license?

   (A) a collection truck which can carry loads up to 11 tons
   (B) an inspector's car that weighs 3495 pounds when empty
   (C) a pickup truck transporting 10,000 pounds of radioactive waste in a lead box
   (D) a spreader with 3 tons of salt and sand to spread

43. The job of the sanitation worker on a collection truck is to collect trash and garbage, not to do the work of a firefighter or police officer. However, the sanitation worker is also a citizen of the community and has certain responsibility to report dangerous situations. Which of the following represents a dangerous situation that a sanitation worker should take the time to report before continuing the collection route?

   (A) Two teenagers are sitting on the front steps of a house sharing a joint.
   (B) Children playing ball on the sidewalk are chasing dropped balls into the street from between parked cars.
   (C) A dog walker is ignoring the dog's droppings instead of scooping them up.
   (D) A wire dangling over the sidewalk is sparking and crackling.

44. While standing on the pier waiting for the collection truck to be unloaded onto a barge, Sanitation Worker Margot Torbert is witness to a hit-and-run accident. She makes mental notes to this effect:

Location: Front Street in front of Pier 32
Victim: very old Asian man
Vehicle: red sports car with out-of-state license plates

Margot Torbert is about to place a telephone call to the local emergency number. Which of the following expresses the information most clearly and accurately?

(A) A very old Asian man was hit by a red sports car at the pier while I watched.
(B) The car that hit the old Asian man on Front Street by Pier 32 left.
(C) I just saw a hit-and-run accident on Front Street at Pier 32. The person hit is an old Asian man and the car was a red out-of-state sports car.
(D) On Front Street down by the garbage piers an old Asian man hit a red sports car with out-of-state plates.

Answer questions 45 to 50 based on the service schedule and the information below.

Department regulations require that all vehicles receive preventive maintenance servicing every 2000 miles or every 3 months, whichever comes first. The schedule below refers to four of the department's vehicles. Today's date is April 8.

Vehicle	Date Last Serviced	Mileage at Service	Current Mileage
A	February 2	13,402	15,306
B	January 13	7,796	8,932
C	February 17	9,841	11,624
D	January 6	36,207	37,810

45. Vehicle A should be serviced

(A) in less than 100 miles
(B) on May 2
(C) today
(D) in exactly 86 miles

46. Vehicle B

(A) need not be serviced for nearly another 1000 miles
(B) should be serviced on the date the mileage reads 9,795
(C) should be serviced within the next five days
(D) is already overdue for service

47. If vehicle C were serviced today, it would be

(A) less than 200 miles before required service
(B) more than a month early
(C) 117 miles before service was due
(D) exactly on schedule

48. Vehicle D should

(A) be serviced in 397 more miles
(B) have been serviced 403 miles ago

(C) be serviced within the next two days

(D) have been serviced two days ago

49. The difference between the current mileage reading on the vehicle that has driven the least miles and the vehicle that has driven the greatest number of miles is

(A) 28,822

(B) 28,878

(C) 29,778

(D) 29,878

50. The vehicle that requires service most urgently is

(A) vehicle A

(B) vehicle B

(C) vehicle C

(D) vehicle D

51. Sanitation Worker McKenzie, covering the morning collection route, notices an apparently unsupervised small boy wandering aimlessly on East 18th Street between Avenue P and Kings Highway. McKenzie asks the child his name and is told "Jimmy." The child, who appears to be no more than three years old, has curly blond hair and is wearing striped pants and a white T-shirt with Bugs Bunny on the front. The youngster cannot tell his last name nor where he lives. McKenzie picks up the nearest emergency phone, identifies himself to the dispatcher, and tells about the child. Which of the following statements expresses this information most clearly and accurately?

(A) Jimmy is all alone on East 18th Street between Avenue P and Kings Highway. He might be missing.

(B) A small blond boy wearing striped pants and a Bugs Bunny T-shirt and identifying himself as "Jimmy" is wandering on East 18th Street between Avenue P and Kings Highway.

(C) I have little Jimmy on East 18th Street between Avenue P and Kings Highway in striped pants. He is blond.

(D) Jimmy is with me at a call box near East 18th Street and Kings Highway. He likes Bugs Bunny and wears striped pants.

Answer questions 52 to 55 based on the licensing requirements for appointment as sanitation worker given below.

1. Must have operator's license and commercial learner's permit
    or
2. Must have Class B Commercial Driver's License
        or, if not already holding a Class 1, 2, or 3 License
3. Must be at least 21 years of age within six months, and
4. Must obtain Class B license within six months of starting work.

52. Marlene Markowitz is 19 years old and holds a Class 2 license. She

(A) must get a commercial learner's permit

(B) must wait until she turns 21

(C) is eligible for immediate appointment

(D) may start work now but must obtain a Class B license within six months

53. Arnold Ie is 23 years old and has never driven a car. He

(A) cannot be a sanitation worker at this time
(B) cannot be appointed until he learns to drive
(C) must get a driver's license and a commercial learner's permit within six months
(D) is too old to get a Class B Commercial Driver's License

54. Bob Adams is 20½ years old and has a driver's license and a commercial learner's permit. He

(A) can start work immediately and earn a Class B Commercial Driver's License within six months
(B) must wait until he turns 21; then he will be fully licensed
(C) cannot become a sanitation worker until he turns 21 and obtains a Class B Commercial Driver's License
(D) can start work on his 21st birthday and then obtain his Class B license within six months

55. All of the following statements are true *except*

(A) You must be 21 years of age to obtain a Class B License.
(B) You may be less than 21 years of age to hold a Class 1 License.
(C) You can hold an operator's license and commercial learner's permit at the same time.
(D) At age 21 a commercial learner's permit becomes a Class B Commercial Driver's License.

56. After a number of low-rainfall seasons, the reservoirs are dangerously low, and the mayor has declared a drought alert. On his way back along the collection route, Charles Brown sees a hydrant wide open and pouring water into the street on 75th Road between 76th Street and 73rd Avenue. Upon Brown's return to the garage, he files a report about the hydrant. Which of the following statements gives the information most clearly and accurately?

(A) Water is all over the street on 76th Street and 73rd Avenue.
(B) An open hydrant is spilling water all over 75th Road between 76th Street and 73rd Avenue.
(C) Water is being wasted by an open hydrant on 75th between 76th and 73rd.
(D) An open hydrant is wasting water on 75th Street between 76th Avenue and 73rd Road.

57.

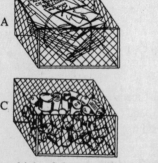

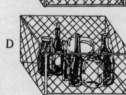

Into which of the recycling bins above should you place styrofoam meat trays?

(A) only in bin A
(B) in bin A and bin D
(C) in any bin
(D) in none of these bins

58. The main reason for the 5¢ deposit on bottles in many areas is

    (A) to raise money for the state
    (B) to give the unemployed a source of income
    (C) to reduce littering
    (D) to avoid the hazards of broken glass on streets and in parks

59. An effect of the supermarket machines that crush cans and bottles has been to

    (A) reduce air pollution
    (B) make more space in landfill dumps
    (C) ruin the refillable bottle industry
    (D) save the lives of marine wildlife

60. Which of the objects below would be most difficult for one person alone to load onto a collection truck?

(A)

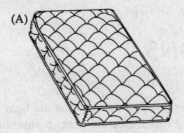

(B)

(C)

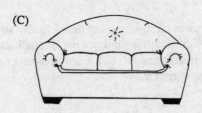

(D)

MODEL EXAM IV ANSWER KEY

1. A	11. D	21. D	31. A	41. B	51. B
2. B	12. C	22. B	32. B	42. B	52. C
3. D	13. B	23. A	33. D	43. D	53. A
4. C	14. A	24. D	34. C	44. C	54. A
5. B	15. A	25. D	35. B	45. A	55. D
6. C	16. D	26. C	36. D	46. C	56. B
7. D	17. B	27. B	37. A	47. B	57. D
8. A	18. C	28. B	38. D	48. D	58. C
9. D	19. C	29. A	39. A	49. B	59. B
10. B	20. B	30. C	40. C	50. D	60. C

EXPLANATIONS

1. **(A)** In a three-color traffic signal, the yellow light serves as a warning that the light is about to change to red. The yellow, then, follows the green. Unless the driver is almost in the intersection, the yellow light warns, "Stop now so that you don't get caught on a red light after entering the intersection." A single flashing yellow light is the general caution signal.

2. **(B)** The diamond-shaped sign is an information sign, not a prohibition sign. The sharply curved arrow means that the road will curve sharply.

3. **(D)** Even if you do not know any first aid, you should be able to answer this type of question using common sense. Turn the person's head so that he or she does not choke on swallowed vomit.

4. **(C)** Dirty snow is no more difficult to cart away than is clean snow, but heavy traffic and parked cars that slow the flow of traffic do make it difficult for plows to get through and for removal equipment to do its job. (D) makes a true statement, but simply describes a situation in the country. It has no effect on city snow removal.

5. **(B)** This is another common sense question.

6. **(C)** Districts 2 and 4 both have 19 flushers.

7. **(D)** District 2 has the smallest number of collection trucks, sweepers and snow plows, and the same number of flushers as District 4. It does not have the fewest number of spreaders. District 4 has the fewest spreaders.

8. **(A)** $72 + 22 + 56 + 30 = 180$

9. **(D)** $312 - 280 = 32$

10. **(B)** District 2 is smallest but District 4 has fewer spreaders. Also, 2 and 4 have the same number of flushers though 4 is larger than 2.

11. **(D)** The arm fully extended out the driver's window is the official hand signal for a left turn.

12. **(C)** The arm pointing downwards is the signal that the driver is about to stop.

13. **(B)** Elbow bent, hand pointing straight up means. "I am about to turn right."

14. **(A)** Carbon monoxide is colorless *and* odorless. It often collects in sewers and is very dangerous. The odor of car exhaust, which contains carbon monoxide, is created by other exhaust gasses, not by the carbon monoxide.

15. **(A)** All of the information would be useful and important, but if the physician does not know where to find the injured person, all other information is totally useless.

16. **(D)** Water flows downhill. The purpose of flushing the street is to clean it, not to spread the dirt. The dirt will flow down the street with the water, leaving a clean area.

17. **(B)** See the first two sentences.

18. **(C)** The fourth sentence discusses the problems of hand injuries when workers avoid wearing gloves.

19. **(C)** Back strain is a problem of sanitation workers, but it is not mentioned in this paragraph.

20. **(B)** The proper clothing is unlikely to prevent accidents, but it should lessen the injuries resulting from those accidents.

21. **(D)** Steel toe-cap inserts protect toes.

22. **(B)** The same traffic that makes it difficult to clear the snow is disrupted by the snow.

23. **(A)** All water is equally effective in cleaning streets. However, reservoir water (drinking water) may be scarce while there is lots of water in the river. River water is often polluted, so it is treated before use on the streets.

24. **(D)** This is a common sense question. Rats live in the garbage.

25. **(D)** Common courtesy never hurts.

26. **(C)** A flashing yellow light always means "caution." The steady, though short, yellow light between green and red means, "Caution, the light is about to turn red."

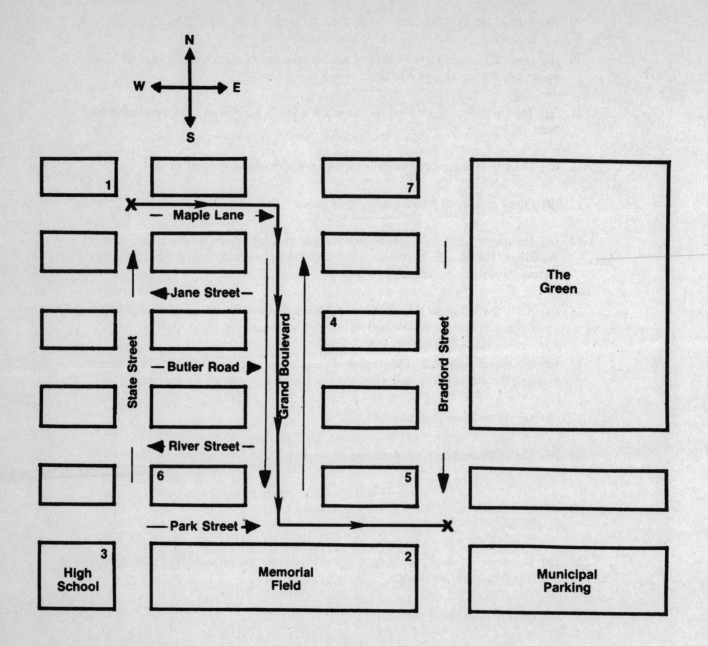

27. **(B)** This is the "school crossing" sign. The two parallel lines on the ground represent a crosswalk; the figures are meant to be school children.

28. **(B)** The sign indicates that you are in the vicinity of a hospital. It would be nice if you would try to be quiet.

29. **(A)** The first sign is the "interstate highway" sign. The second (B) indicates an older federal highway not designated as an *interstate* even though it goes from one state to another. (C) designates a state highway. (D) is a "stop" sign.

30. **(C)** See the first sentence.

31. **(A)** $12 - 2\frac{1}{2} = 9\frac{1}{2}$

32. **(B)** The first sentence says that the new trucks are larger. We have no way of knowing from this passage whether or not they are also cleaner or more durable. Being larger, the new trucks are likely to be more difficult to drive.

33. **(D)** The second sentence of the second paragraph tells us that heavy cans of coal ash were lifted and emptied into the truck. Only when it caught on fire was coal ash dumped, watered down, and reloaded by shovel.

34. **(C)** See the last sentence of the first paragraph.

35. **(B)** *Frequent* means *occurring often*. Thus, fires in the garbage trucks were a common occurrence. Undoubtedly, they were time consuming as well, but that is not the meaning of *frequent*.

36. **(D)** This statement tells what and where. (A) does not say where on Grand Street. (Grand Street could be very long.) (B) is unclear on location and is written as if Daniels was not able to clean the car. (C) gives no idea of how long the car has been there nor does it imply that it presents a problem for street cleaning.

37. **(A)** Step 2 tells to pull the vehicle out of the flow of traffic, if possible. But we know that the accident has occurred in a narrow street lined with parked cars on both sides. Since it is not possible to get out of the flow of traffic, Cabrini should summon a police officer.

38. **(D)** The correct route is mapped out on page 106. If you turn the page so as to face the way the vehicle is facing at each turn, you will find it easier to keep right and left straight.

39. **(A)** Look at the map on page 108 to see how easy this route is. Choice (B) is a legal route, but it is unnecessarily long. Choice (C) is just as short as (A), but it sends you the wrong way on Bradford. (D) is both longer than necessary and sends you the wrong way on Park.

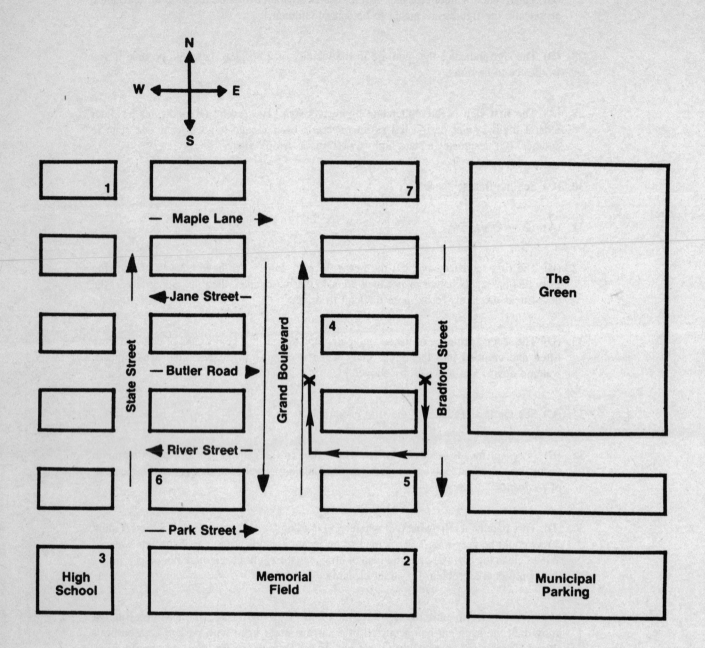

40. **(C)** As you can see on the map below, this route is direct and legal. Choice (A) sends you the wrong way on Bradford. Choice (B) is longer, involves extra turns, and goes the wrong way on Bradford. (D) is second best. This route is legal and not too long, but it involves an extra turn. The straightest route is best.

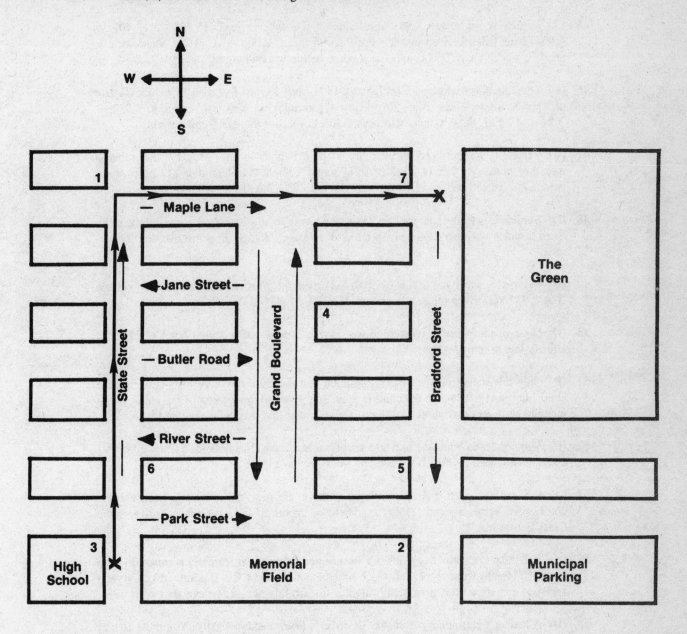

41. **(B)** The quotation speaks of commercial driver's licenses and then says, "....the new licenses which are necessary to drive Sanitation trucks." The new rule is that you must hold a commercial driver's license to drive a sanitation truck. The paragraph speaks of possible future shortages of drivers, not that there is current shortage. Not everyone who passes a driving test is a good driver, though that would be an ideal situation.

42. **(B)** It would be impossible to load 22,500 pounds of cargo into a car. (A) and (D) exceed the weight limits and must be driven by licensed commercial drivers. (C) carries hazardous materials so its driver must have a commercial license regardless of what the truck weighs.

43. **(D)** A live wire is dangerous and should be reported at once.

44. **(C)** This report gives an accurate location and tells what happened. (A) is not complete as to location. (B) does not describe the hit-and-run vehicle. (D) has the facts wrong.

45. **(A)** Vehicle A has driven 1,904 miles since it was last serviced. $15,306 - 13,402 = 1,904$. It must be serviced in not more than 96 miles. $2,000 - 1,904 = 96$. May 2 is nearly a month away; 2,000 miles will come before 3 months pass.

46. **(C)** Three months from service on January 13 is April 13, just five days from today's date of April 8. Since vehicle B has driven just 1,136 miles since its last servicing, $8,932 - 7,796 = 1,136$, three months will pass before it reaches the 2,000 mile mark.

47. **(B)** Vehicle C was serviced on February 17 so it is not due until May 17, more than a month from today's date of April 8. In mileage, Vehicle C has another 217 miles to go before it requires service. $11,624 - 9,841 = 1,783$ $2,000 - 1,783 = 217$.

48. **(D)** Vehicle D was last serviced on January 6, so it should have been serviced again by April 6 which was two days ago. In terms of mileage, Vehicle D is not overdue for service.

49. **(B)** Vehicle D, which has driven 37,810, has driven the greatest number of miles. Vehicle B at 8,932 has driven the fewest miles. $37,810 - 8,932 = 28,878$.

50. **(D)** None of the vehicles is overdue for service in terms of mileage, but Vehicle D is already late in terms of time.

51. **(B)** Only this statement gives the information that the child might be lost, describes the child fully, and tells where the child is now. (A) gives no description of the child; (C) is incomplete in its description; (D) doesn't give a clue that this child may be lost.

52. **(C)** Marlene Markowitz falls into the exception. Since she already holds a Class 2 license, she does not need a Class B license and does not have to be 21.

53. **(A)** It is not enough that Arnold Ie learn to drive. He must get a driver's license and a commercial learner's permit. Only then can he be appointed and have 6 months to earn the Class B license.

54. **(A)** Bob Adams already has a driver's license and commercial learner's permit. Since he is only 6 months from age 21, at which time he could get a Class B commercial Driver's License, and since he is permitted 6 months to earn this license, he can start work now.

55. **(D)** A learner's permit never turns into a driver's license automatically. You must always pass the test. You must be at least 21 to be issued a Class B Commercial Driver's License.

56. **(B)** Read carefully. The other choices are inaccurate in location description.

57. **(D)** Styrofoam does not fit into any of these categories. Actually there are few recycling uses for styrofoam, so you are not likely to find styrofoam separated for recycling.

58. **(C)** Authorities hope that people will return their bottles to get back the 5¢ deposit. If people bring back their bottles, the bottles will not be left lying around and littering the ground.

59. **(B)** Crushed bottles and cans take up much less space than uncrushed empties.

60. **(C)** One person working alone would have the hardest time with the big, bulky sofa. The carpeting could be hoisted over one shoulder and carried for a short distance.

ANSWER SHEET FOR MODEL EXAM V

1. Ⓐ Ⓑ Ⓒ Ⓓ   16. Ⓐ Ⓑ Ⓒ Ⓓ   31. Ⓐ Ⓑ Ⓒ Ⓓ   46. Ⓐ Ⓑ Ⓒ Ⓓ

2. Ⓐ Ⓑ Ⓒ Ⓓ   17. Ⓐ Ⓑ Ⓒ Ⓓ   32. Ⓐ Ⓑ Ⓒ Ⓓ   47. Ⓐ Ⓑ Ⓒ Ⓓ

3. Ⓐ Ⓑ Ⓒ Ⓓ   18. Ⓐ Ⓑ Ⓒ Ⓓ   33. Ⓐ Ⓑ Ⓒ Ⓓ   48. Ⓐ Ⓑ Ⓒ Ⓓ

4. Ⓐ Ⓑ Ⓒ Ⓓ   19. Ⓐ Ⓑ Ⓒ Ⓓ   34. Ⓐ Ⓑ Ⓒ Ⓓ   49. Ⓐ Ⓑ Ⓒ Ⓓ

5. Ⓐ Ⓑ Ⓒ Ⓓ   20. Ⓐ Ⓑ Ⓒ Ⓓ   35. Ⓐ Ⓑ Ⓒ Ⓓ   50. Ⓐ Ⓑ Ⓒ Ⓓ

6. Ⓐ Ⓑ Ⓒ Ⓓ   21. Ⓐ Ⓑ Ⓒ Ⓓ   36. Ⓐ Ⓑ Ⓒ Ⓓ   51. Ⓐ Ⓑ Ⓒ Ⓓ

7. Ⓐ Ⓑ Ⓒ Ⓓ   22. Ⓐ Ⓑ Ⓒ Ⓓ   37. Ⓐ Ⓑ Ⓒ Ⓓ   52. Ⓐ Ⓑ Ⓒ Ⓓ

8. Ⓐ Ⓑ Ⓒ Ⓓ   23. Ⓐ Ⓑ Ⓒ Ⓓ   38. Ⓐ Ⓑ Ⓒ Ⓓ   53. Ⓐ Ⓑ Ⓒ Ⓓ

9. Ⓐ Ⓑ Ⓒ Ⓓ   24. Ⓐ Ⓑ Ⓒ Ⓓ   39. Ⓐ Ⓑ Ⓒ Ⓓ   54. Ⓐ Ⓑ Ⓒ Ⓓ

10. Ⓐ Ⓑ Ⓒ Ⓓ   25. Ⓐ Ⓑ Ⓒ Ⓓ   40. Ⓐ Ⓑ Ⓒ Ⓓ   55. Ⓐ Ⓑ Ⓒ Ⓓ

11. Ⓐ Ⓑ Ⓒ Ⓓ   26. Ⓐ Ⓑ Ⓒ Ⓓ   41. Ⓐ Ⓑ Ⓒ Ⓓ   56. Ⓐ Ⓑ Ⓒ Ⓓ

12. Ⓐ Ⓑ Ⓒ Ⓓ   27. Ⓐ Ⓑ Ⓒ Ⓓ   42. Ⓐ Ⓑ Ⓒ Ⓓ   57. Ⓐ Ⓑ Ⓒ Ⓓ

13. Ⓐ Ⓑ Ⓒ Ⓓ   28. Ⓐ Ⓑ Ⓒ Ⓓ   43. Ⓐ Ⓑ Ⓒ Ⓓ   58. Ⓐ Ⓑ Ⓒ Ⓓ

14. Ⓐ Ⓑ Ⓒ Ⓓ   29. Ⓐ Ⓑ Ⓒ Ⓓ   44. Ⓐ Ⓑ Ⓒ Ⓓ   59. Ⓐ Ⓑ Ⓒ Ⓓ

15. Ⓐ Ⓑ Ⓒ Ⓓ   30. Ⓐ Ⓑ Ⓒ Ⓓ   45. Ⓐ Ⓑ Ⓒ Ⓓ   60. Ⓐ Ⓑ Ⓒ Ⓓ

Number Right _____

MODEL EXAM V

60 QUESTIONS—2½ HOURS

Directions: Read each question and choose the best answer. On your answer sheet, darken the letter of the answer you choose.

Answer questions 1 through 3 only on the basis of the following passage.

"Helping to prevent accidents is the job of every worker. Tell your foreman about unsafe equipment right away. Wear safe clothing. Bend your knees when lifting and get somebody to help you with very heavy objects."

1. The above passage says that helping to prevent accidents is the job of

   (A) the foreman
   (B) the safety division
   (C) management
   (D) every worker

2. Equipment that is not safe should be

   (A) used with special care
   (B) reported to your foreman right away
   (C) marked with a red tag
   (D) parked at the side of the road

3. When lifting very heavy objects, you should

   (A) ask your foreman what to do
   (B) keep your legs straight
   (C) always wear protective gloves
   (D) get somebody to help you

4. "Sanitation workers are expected to do the work they are assigned. In addition, however, they are expected to have good relations with the public in order to reduce the likelihood of complaints." According to this statement, good relations with the public are desirable because

   (A) they may reduce the number of complaints
   (B) sanitation workers are highly qualified

   (C) a worker's personality is important on the job
   (D) public relations is the most important part of the job of sanitation worker

5. "People who do not work with their hands do not know how hard it is to do manual work." According to this statement, manual work is work which is

   (A) not hard to do
   (B) done by many people
   (C) done with the hands
   (D) understood by few people

Answer questions 6 and 7 only on the basis of the following passage:

"The Department of Sanitation starts early in May to prepare for snow expected during the following winter. It begins by fixing the snow removal equipment which was used during the winter. It is then usually kept busy with either snow removal or preparation for snow removal every month through the end of March."

6. According to the above passage, for how many months during the year is the Department of Sanitation busy with either snow removal or preparation for snow removal?

   (A) 9   (B) 10   (C) 11   (D) 12

7. According to the passage, in the month of May the Department of Sanitation

   (A) stores the snow removal equipment
   (B) fixes the snow removal equipment
   (C) equips the sanitation workers
   (D) collects the garbage piled up because of snow

Answer questions 8 and 9 only on the basis of the following passage:

"It would be unusual for a snow storm to develop without warning. When a warning is received, sanitation workers load the salt spreaders and attach plows to the trucks."

8. According to the passage, a snow storm seldom develops

  (A) without advance notice
  (B) in the spring
  (C) in New York City
  (D) without rain coming first

9. Once a snow warning is received, sanitation workers prepare for the storm by

  (A) removing plows from the trucks
  (B) greasing and oiling the salt spreaders
  (C) emptying the salt spreaders
  (D) putting plows on the truck

10. "Because trucks are heavier than cars, they are more difficult to slow down and stop." According to this statement, the reason trucks are more difficult to stop than cars is that

  (A) cars are slower than trucks
  (B) cars usually have automatic shift
  (C) trucks are more likely to stall
  (D) trucks weigh more than cars

Questions 11 through 20 test your understanding of the vocabulary in each question.

11. "The large *vehicle* was being repaired." Which of the following is a *vehicle*?

  (A) truck   (C) boiler
  (B) building   (D) table

12. "The *fence* needs to be painted." The one of the following which is most like a *fence* is a

  (A) door   (C) wall
  (B) crane   (D) building

13. "*Furniture* is not taken with the regular collection." Which of the following is *furniture*?

  (A) sofas and chairs
  (B) cars and trucks
  (C) brooms and mops
  (D) bags and boxes

14. "The *group* was assigned to do special work." Which of the following is a *group*?

  (A) truck   (C) team
  (B) boat   (D) foreman

15. "Sanitation workers often use *tools* in their work." The one of the following which is most often considered a *tool* is a

  (A) tire   (C) glove
  (B) shovel   (D) basket

16. "The man claimed that he could not *lift* the can." The word *lift* means most nearly

  (A) bury   (C) refill
  (B) pick up   (D) clean

17. "Place all the boxes *below* the second shelf." The word *below* means

  (A) under   (C) beside
  (B) into   (D) over

18. "The truck could not go under the bridge because the bridge was too *low*." The reason the truck could not go under the bridge was that the bridge was

  (A) not high enough
  (B) not long enough
  (C) not strong enough
  (D) not wide enough

19. "He could not get his truck on the *highway*." A *highway* is a type of

  (A) lot   (C) scale
  (B) road   (D) sidewalk

20. "This street should be *clean* when the sanitation workers finish." The word *clean* means free of

  (A) obstacles   (C) traffic
  (B) pedestrians (D) dirt

21. "Even minor injuries on the job should be reported to one's supervisor and should be treated by a doctor if necessary." According to this statement all injuries on the job

  (A) are usually minor
  (B) should be reported to one's supervisor

(C) should be treated by a doctor
(D) can be avoided

22. ''Up to the age of forty, people in good health should have a medical check-up once a year. People over 40 should have a check-up twice a year. People in poor health should have a check-up even more frequently.'' According to this statement, people under 40 who are in good health should have a medical check-up

(A) once a year
(B) twice a year
(C) more frequently than twice a year
(D) once every other year

23. ''After stopping a sanitation truck, it is very important that you look out for children just before stepping on the gas pedal again.'' According to this statement, it is necessary to look out for children

(A) after putting a sanitation truck in motion
(B) immediately after stopping a sanitation truck
(C) immediately before stepping on the gas pedal
(D) before stopping a sanitation truck

Above each of questions 24 through 29 there is a sign followed by a question and four choices. Look at the sign and then blacken the space on your answer sheet next to the letter that best describes the meaning of the sign.

24. This sign tells you to

(A) give up the right-of-way ahead
(B) turn right ahead
(C) increase your speed ahead
(D) park your vehicle ahead

25. This sign tells you that when the road is wet

(A) your vehicle may stall
(B) your vehicle may slide
(C) the pavement may crack
(D) you may not use this road

26. This sign tells you that, about 1000 feet ahead, the street

(A) is dangerous
(B) curves to the right
(C) is not open
(D) becomes a highway

REDUCED
SPEED
AHEAD

27. This sign tells you that further ahead you must

(A) go slower
(B) go faster
(C) continue at the same speed
(D) make a left turn

NO
TURNS

28. This sign tells you that you must

(A) go straight ahead
(B) stay in your lane
(C) not pass another car
(D) not slow down

DO NOT
ENTER

29. This sign tells you that you must *not*

(A) go in
(B) speed up
(C) slow down
(D) stop

In questions 30 through 39 pick the answer which means most nearly the same as the word in the sentence.

30. "The sanitation workers *combined* the contents of the two boxes." The word *combined* means

(A) sifted through   (C) tore apart
(B) put together   (D) forgot about

31. "Don't touch the *lever* on the left side." The word *lever* means

(A) button   (C) handle
(B) rope   (D) gun

32. "All *litter* should be taken away." The word *litter* means

(A) paint   (C) rubbish
(B) bowls   (D) evidence

33. "The *inspection* of the street was complete." The word *inspection* means

(A) cleaning   (C) repair
(B) examination   (D) painting

34. "The *route* must be followed exactly." The word *route* means

(A) foreman   (C) way
(B) truck   (D) recipe

35. "Don't *injure* your back." The word *injure* means

(A) bend   (C) hurt
(B) use   (D) exercise

36. "John *repaired* the machine. The word *repaired* means

(A) fixed   (C) ran
(B) broke   (D) oiled

37. "Put the *lid* on the box." The word *lid* means

(A) cover   (C) rope
(B) ribbon   (D) wrapping

38. "The *rear* of the truck should be washed." The word *rear* means

(A) hood   (C) back
(B) front   (D) roof

39. "Sanitation workers must *assist* each other while at work." The word *assist* means

(A) help   (C) like
(B) outdo   (D) hurt

Answer questions *40* through *45* by performing the operation required (addition or subtraction).

40. Add: 10,487
           + 145

(A) 10,342   (C) 10,632
(B) 10,622   (D) 10,652

41. Add: 26,836
           + 87

(A) 26,749   (C) 26,943
(B) 26,923   (D) 26,973

42. Subtract: 83,204
               − 83,075

(A) 109   (B) 129   (C) 139   (D) 144

43. Subtract: 19,095
            − 19,029

 (A) 66   (B) 74   (C) 79   (D) 86

44. If the mileage indicator on your truck reads 14,382 at the beginning of the day, and it reads 14,431 at the end of the day, the number of miles that the truck has been driven that day is

 (A) 29   (B) 34   (C) 39   (D) 49

45. On a certain day your truck makes three trips to the dumping area and dumps 5.5 tons, 6.3 tons, and 4.8 tons of trash. The total number of tons of trash that your truck has dumped that day is

 (A) 15.0   (B) 15.6   (C) 16.0   (D) 16.6

Assume that, while driving a Department of Sanitation vehicle, you have been involved in an accident and that you are required to fill out an Accident Report form.

Answer questions 46 through 54 based on the Accident Report form shown on the next page.

46. The accident took place at the corner of Jerome Avenue and 170th Street. On the form, you should write "the corner of Jerome Avenue and 170th St." in the block entitled

 (A) LOCATION OF ACCIDENT
 (B) TYPE OF ACCIDENT
 (C) WEATHER
 (D) LIGHT CONDITIONS

47. The accident happened when the sides of both vehicles came into contact as the other driver was trying to pass you. On the form, in the block entitled TYPE OF ACCIDENT, you should check the box marked

 (A) Head On   (C) Right Angle
 (B) Sideswipe   (D) Front End

48. At the location of the accident a police officer was directing traffic. The proper place on the form for this information is the block entitled

 (A) TYPE OF TRAFFIC CONTROL NEAREST TO LOCATION OF ACCIDENT
 (B) TYPE OF ACCIDENT

(C) JOB BEING PERFORMED AT TIME OF ACCIDENT
(D) ROAD SURFACE AND CONDITION.

49. Since the accident happened at 11:00 P.M., you should write "11:00 P.M." in Block #

 (A) 1   (B) 3   (C) 5   (D) 7

50. On the street where the accident happened, all traffic had to go in the same direction. On the form, in the block entitled ROAD TYPE, you should check the box next to

 (A) Not Separated   (C) Separated
 (B) Two-way   (D) One-way

51. When the accident happened, it was raining and you were on a black-top street. On the form, in the block entitled ROAD SURFACE and CONDITIONS, you should check the two boxes next to

 (A) Wet and Concrete
 (B) Dry and Concrete
 (C) Wet and Black-top
 (D) Dry and Black-top

52. When the accident happened, you were driving along your collection route. You should put this information in the block entitled

 (A) ROAD SURFACE AND CONDITION
 (B) LOCATION OF ACCIDENT
 (C) TYPE OF TRAFFIC CONTROL NEAREST TO THE LOCATION
 (D) JOB BEING PERFORMED AT TIME OF ACCIDENT

53. The accident happened on March 2, 19--. On the form, you should write "March 2, 19--" in Block #

 (A) 1   (B) 4   (C) 7   (D) 10

54. The name of the other person involved in the accident is Frank Smith. On the form, you should put this information in the block entitled

 (A) JOB BEING PERFORMED AT TIME OF ACCIDENT
 (B) LOCATION OF ACCIDENT
 (C) TYPE OF TRAFFIC CONTROL NEAREST TO LOCATION
 (D) OTHER VEHICLE

ACCIDENT REPORT

Block #1 DATE AND TIME

Date of Accident _____

Time of Accident _____

Block #2 LOCATION OF ACCIDENT

_____

_____

Block #3 DEPARTMENT OF SANITATION VEHICLE

Operator's Name _____

Operator's License # _____

Vehicle License # _____

Block #4 OTHER VEHICLE

Vehicle License # _____

Owner's Name _____

Operator's Name _____

Operator's License # _____

Block #5 TYPE OF ACCIDENT

Head On ☐   Front End ☐

Side Swipe ☐   Non Collision ☐

Right Angle ☐

Block #6 LIGHT CONDITION

Daylight ☐   Dusk ☐

Dark ☐   Dawn ☐

Block #7 TYPE OF TRAFFIC CONTROL NEAREST TO LOCATION OF ACCIDENT

Police Officer ☐   Stop Sign ☐

Crossing Guard ☐   Yield Sign ☐

Traffic Light ☐

Block #8 ROAD TYPE

One-Way ☐   Separated ☐

Two-Way ☐   Not Separated ☐

Block #9 WEATHER CONDITIONS

Clear ☐   Sleeting ☐   Snowing ☐

Raining ☐   Fog ☐

Block #10 ROAD SURFACE AND CONDITION

Black-top ☐   Wet ☐

Concrete ☐   Dry ☐

Block #11   JOB BEING PERFORMED AT TIME OF ACCIDENT

Driving Along Collection Route ☐   Stopped: Loading Along Route ☐

Driving to Dumping Area ☐   Other _____ ☐

Answer questions 55 through 57 only on the basis of the following passage:

"A heavy snowfall may cause delays in the movement of trains and buses. People are often late for work when it snows. Both pedestrians and cars have accidents because of snow and ice. Pedestrians slip and fall. Cars skid and collide."

55. The above passage indicates that heavy snow

   (A) is a beautiful thing to see
   (B) may make the trains run late
   (C) gives temporary work to the unemployed
   (D) should be cleared from sidewalks within four hours

56. According to the above passage, snow and ice may cause cars to

   (A) slow down   (C) stall
   (B) freeze   (D) skid

57. The above passage says that, when it snows

   (A) children love to have snowball fights
   (B) people are often late for work
   (C) garbage collection is halted
   (D) snow plows must be attached to garbage trucks

Answer questions 58 through 60 based only on the information in the passage below.

"Sanitation workers sometimes have to listen to complaints from the public. When an angry citizen complains to you, you should remember to stay calm. If you can answer the complaint, you should do so. If you cannot answer the complaint, you should refer the citizen to someone who can answer it."

58. Sanitation workers who come into contact with the public sometimes have to

   (A) sweep up trash
   (B) shout at citizens
   (C) listen to complaints
   (D) help put out fires

59. If a citizen who is complaining to you is very angry, you should

   (A) get angry
   (B) stay calm
   (C) ignore him
   (D) tell him to leave you alone

60. If you cannot answer the complaint you should

   (A) make up something that sounds logical
   (B) ask a passer-by for the information
   (C) tell him who can give him the answer
   (D) tell him you do not know and walk away

MODEL EXAM V ANSWER KEY

1. D	11. A	21. B	31. C	41. B	51. C
2. B	12. C	22. A	32. C	42. B	52. D
3. D	13. A	23. C	33. B	43. A	53. A
4. A	14. C	24. A	34. C	44. D	54. D
5. C	15. B	25. B	35. C	45. D	55. B
6. C	16. B	26. C	36. A	46. A	56. D
7. B	17. A	27. A	37. A	47. B	57. B
8. A	18. A	28. A	38. C	48. A	58. C
9. D	19. B	29. A	39. A	49. A	59. B
10. D	20. D	30. B	40. C	50. D	60. C

EXPLANATIONS

1. **(D)** This is exactly what the first sentence says.

2. **(B)** See the second sentence.

3. **(D)** You should not lift heavy objects alone; get help and bend your knees when lifting.

4. **(A)** Good public relations help to reduce complaints. The most important part of the job of a sanitation worker is to do the assigned work.

5. **(C)** If people who do not work with their hands do not know how hard it is to do manual work, then manual work must be done with the hands.

6. **(C)** From the beginning of May until the end of March is 11 months. Only in the month of April is snow removal of no concern.

7. **(B)** See the second sentence.

8. **(A)** Refer to the first sentence.

9. **(D)** Preparations for a snowstorm involve putting the plows on the trucks and loading the salt spreaders.

10. **(D)** The weight of trucks makes them more difficult to slow down and stop than cars.

11. **(A)** A *vehicle* is a device for transporting people or objects. A truck is a vehicle.

12. **(C)** A wall serves the same function as a *fence*—dividing one area from another.

13. **(A)** *Furniture* is the movable equipment in a room which equips it for living.

14. **(C)** A *group* consists of more than one.

15. **(B)** A *tool* is an implement used for cutting, hitting, digging, rubbing, etc.

16. **(B)** To *lift* is to raise to a higher position, hence to pick up.

17. **(A)** *Below* means in a lower place or under.

18. **(A)** In this sentence, *too low* means not high enough.

19. **(B)** A *highway* is a public roadway.

20. **(D)** That which is *clean* is not dirty. The word to be used in getting rid of obstacles, pedestrians or traffic is "clear."

21. **(B)** The certain statement is that all injuries on the job should be reported. If *even* minor injuries should be reported, obviously some injuries are more serious. Treatment by a doctor is not necessarily always required.

22. **(A)** "Up to the age of forty" means under forty. These are the people for whom a check-up once a year is sufficient.

23. **(C)** The sentence says that you should look out for children after you have been stopped and before restarting.

24. **(A)** To *yield* is to let someone else go first.

25. **(B)** Your vehicle may slide on a *slippery*, wet road.

26. **(C)** If the street is *closed*, it is not open.

27. **(A)** To *reduce* is to lessen. When you lessen your speed, you go slower.

28. **(A)** If you cannot turn, you must go straight ahead.

29. **(A)** *Do not enter* means "stay out!" In other words, do not go in.

30. **(B)** To *combine* is to put together.

31. **(C)** A *lever* is a prying bar or a handle.

32. **(C)** As used in this sentence, *litter* refers to things lying around in disorder or scraps of rubbish.

33. **(B)** *Inspection* is critical examination.

34. **(C)** A *route* is a set course to be travelled.

35. **(C)** To *injure* is to do physical harm to or to hurt.

36. **(A)** To *repair* is to put back into good condition after damage, in other words, to fix.

37. **(A)** A *lid* is a cover.

38. **(C)** The *rear* is the last part or the back.

39. **(A)** To *assist* is to help.

40. **(C)** (Practice doing addition and subtraction accurately without using a calculator. Calculators are almost never permitted at exams. You will either be issued scrap paper or will be

told to use test-booklet margins for your calculations. You do not have to do the arithmetic in your head.)

41. **(B)**

42. **(B)**

43. **(A)**

44. **(D)** 14,431 − 14,382 = 49

45. **(D)** 5.5 + 6.3 + 4.8 = 16.6

46. **(A)** The place where the accident took place is the location of the accident. This information goes into Block #2.

47. **(B)** An accident that occurs when the sides of vehicles come into contact is a "sideswipe." This information is to be entered in Block #5.

48. **(A)** A police officer was controlling traffic near the location of the accident. This information goes into Block #7. The entry concerning "job being performed at time of accident" refers to the job being done by the sanitation worker, not by the police officer.

49. **(A)** The time of the accident goes in Block #1.

50. **(D)** If all traffic must move in the same direction, it is a one-way street. This information belongs in Block #8.

51. **(C)** If it was raining, we can assume that the black-top road surface was wet. The entry goes into Block #10.

52. **(D)** Now we get to the job that the sanitation worker was doing. The entry goes into Block #11.

53. **(A)** The date of the accident goes into Block #1.

54. **(D)** Block #4 requests information about the other vehicle in the accident and its driver and owner.

55. **(B)** A heavy snow that delays the movement of trains and buses may make the trains run late. All the other choices may be true, but they are not indicated by the passage.

56. **(D)** See the last sentence.

57. **(B)** This answer is in the second sentence.

58. **(C)** This is what the first sentence says.

59. **(B)** This advice is given in the second sentence.

60. **(C)** "Refer the citizen to someone who can answer it" means "tell him who can give him the answer."

ACCIDENT REPORT

COMPLETED

Block #1 DATE AND TIME

Date of Accident March 2, 19- -

Time of Accident 11:00 P.M.

Block #2 LOCATION OF ACCIDENT

corner of Jerome Avenue and 170th St.

Block #3 DEPARTMENT OF SANITATION VEHICLE

Operator's Name _____

Operator's License # _____

Vehicle License # _____

Block #4 OTHER VEHICLE

Vehicle License # _____

Owner's Name _____

Operator's Name Frank Smith

Operator's License # _____

Block #5 TYPE OF ACCIDENT

Head On	☐	Front End	☐
Side Swipe	☒	Non Collision	☐
Right Angle	☐		

Block #6 LIGHT CONDITION

Daylight	☐	Dusk	☐
Dark	☒	Dawn	☐

Block #7 TYPE OF TRAFFIC CONTROL NEAREST TO LOCATION OF ACCIDENT

Police Officer	☒	Stop Sign	☐
Crossing Guard	☐	Yield Sign	☐
Traffic Light	☐		

Block #8 ROAD TYPE

One-Way	☒	Separated	☐
Two-Way	☐	Not Separated	☐

Block #9 WEATHER CONDITIONS

Clear	☐	Sleeting	☐	Snowing	☐
Raining	☒	Fog	☐		

Block #10 ROAD SURFACE AND CONDITION

Black-top	☒	Wet	☒
Concrete	☐	Dry	☐

Block #11 JOB BEING PERFORMED AT TIME OF ACCIDENT

Driving Along Collection Route	☒	Stopped: Loading Along Route	☐
Driving to Dumping Area	☐	Other _____	☐

PART THREE

PHYSICAL REQUIREMENTS

MEDICAL STANDARDS

The duties of a sanitation worker are physically demanding. The sanitation worker must lift, carry, dump, throw, and shovel along with driving and operating machinery. The Sanitation Department wants to hire only individuals who can perform this heavy-duty work day-in and day-out without running out of energy and without damaging their health. Further, the Department wants employees who can remain alert to street hazards and traffic even while carrying on physical work.

This is a tall order. Most sanitation departments will require that candidates submit to a medical examination by department doctors or by doctors designated by the Department. Some will permit candidates to be examined by their own physicians, who must follow a departmental procedure in giving the examination. Either way, the candidates must meet up to many standards in order to be appointed.

As with all other requirements, there are variations in medical standards from locality to locality. Each department can decide what is important and what it feels it can overlook. The following list of standards is long and quite inclusive. Your own city or town may have similar standards, or your standards could be stricter or more lenient.

Weight: Overweight to a significant degree which will impair ability to perform the duties of the position rejects. Severe underweight also rejects. Body frame and build are to be considered.

Height: Height of less than 5 feet 3 inches (bare feet) rejects.

Vision: Vision of less than 20/40 both eyes together (eyeglasses allowed) rejects. Peripheral visual fields of less than 140 degrees, rejects. Defective color vision, with emphasis on red, green, and amber, rejects.

Hearing: Inability to hear a conversational voice (hearing aid **not** allowed) rejects.

Other rejecting conditions: hernia; epilepsy; paralysis; heart condition; severe allergies; varicose veins; ulcer; hypertension; diabetes; any significant respiratory condition; chronic alcoholism; chronic drug abuse.

Also rejecting are conditions which interfere with movement or with use of the arms and legs, such as amputation; inequality in size or length of extremities; spina bifida; spondylolisthesis; and other back conditions.

A candidate may also be rejected for a history of mental or nervous ailment. In fact, the medical examiners can reject on the basis of any disease, injury, condition or abnormality which in their opinion would prevent performance of duties or represent a danger to the health of the individual.

You must be healthy to be a sanitation worker.

THE PHYSICAL PERFORMANCE TEST

The job of the sanitation worker is physically stressful. The work involves heavy lifting and carrying while the worker must remain alert and aware of traffic hazards and other personal dangers. Because the physical ability of the sanitation worker is so important to job performance, the final hiring decision is based upon the physical status of the applicant. You must pass the written test to be eligible to take the physical performance test; but you must get a high score on the physical performance test to get the job. Since your rank on the hiring list depends so heavily on your physical condition, it makes sense to devote as much attention to preparing your body for the physical test as to preparing your mind for the written exam.

The first Physical Performance test, described in the next pages, was recently administered to candidates for the position of sanitation worker in the city of New York. Details of the examination will vary from city to city and from locality to locality, but in general all sanitation workers must lift and carry over a sustained period of time. All jurisdictions measure the applicant's ability to do this work efficiently and quickly without doing damage to his or her health. Your exam may not be identical to this but it will be similar in many ways.

New York City advised candidates to report to the physical test site by appointment wearing comfortable work clothes and nonskid work shoes or sneakers. At the test site, applicants watched an orientation video so that they would understand the requirements of the tasks at hand and were issued protective work gloves to wear during the test.

Candidates were told: "Speed of performance enters into the scoring of the physical performance test, though there are no announced time limits of the various events. A time clock will be used to record how long each candidate takes to complete each event. Fast time is crucial to a competitive rank on the list."

This physical performance test was described as a series of five events some of which included a number of separate tasks.

First Physical Performance Test

Event I: Ladder Climb

This event is meant to simulate the climbing of a salt spreader. The ladder is about 10 feet in height. You will stand at the start line approximately two feet from the ladder. At the examiner's signal, you will climb the permanently secured ladder onto the platform, walk across the platform, touch the back railing with at least one hand, walk back to the ladder, and descend the ladder to the floor. You must put at least one foot on each rung both going up and going down. You must face the ladder both going up and going down. You may not jump from the ladder. The examiner will start the clock for Event I at the start signal and stop the clock when either foot touches the ground. The examiner will then direct you to Event II.

Event II: Basket Retrieval

You will stand at the start line approximately two feet from an empty street-corner basket. At the examiner's signal, you will drag the empty basket, which weighs about 30 pounds, a distance of eight feet, step over the line on the floor with at least one foot, lift the basket, place it on its side into a simulated sanitation truck, remove both hands from the basket, retrieve the basket, and drag the basket back to the starting position. The examiner will start the clock at the start signal and stop the clock for Event II when you have returned the basket to its starting position. The examiner will then direct you to Event III.

Event III: Emptying Street-Corner Baskets

Event III consists of two tasks which must be performed consecutively. In the first task you will stand at the start line, approximately two feet from the first basket. At the examiner's signal, the clock will be started. You will drag the first of the four filled street baskets eight feet to a sanitation truck, step over the line on the floor with at least one foot, lift the basket, and empty the contents into the truck. You will then drag the basket back to its starting position. When it is returned, you will drag the next three baskets, one at a time, to the truck, empty each, and bring each back to its starting position. Each basket weighs about 30 pounds and the bags in the baskets weigh 3, 5, 15, and 13 pounds. Next, you will drag the black basket filled with a 35-pound bag to the truck. Do *not* lift the basket to empty it. You must remove the 35-pound bag from the basket and put it into the truck. You may either leave the basket upright or place the basket on its side before removing the bag from the basket and lifting the bag into the truck. The total weight you will handle in this task of Event III is approximately 225 pounds.

You will immediately proceed to the second task in Event III. You will drag the first of the four filled baskets to a truck by going around a large wooden box simulating a car (you must follow the route marked on the floor when going around the car), step over the line on the floor with at least one foot, lift the basket, and empty the contents into the truck. Following the same route, you will drag the basket back to its starting position. Now you will drag the next three baskets one at a time to the truck and, following the same route, bring each basket back to its starting position. You must follow the route marked on the floor when going around the car. Each basket weighs about 30 pounds, and the bags weigh 3, 5, 13, and 25 pounds. Next you will drag the black basket filled with a 33-pound bag to the truck. Do not lift the basket to empty it. You must remove the 33-pound bag from the basket and put it into the truck. You may either leave the basket upright or place the basket on its side before removing the bag from the basket and lifting the bag into the truck. The total weight you will handle in the second task of Event III is approximately 230 pounds. The examiner will stop the clock for Event III when you return the last basket to its starting position. The examiner will then direct you to the rest area where you will rest for a mandatory two minutes before proceeding to Event IV. This two-minute rest will not be counted against your time. It is for your own protection.

Event IV: Carry or Drag Garbage Around Obstacles

Event IV consists of five tasks which must be performed consecutively. You will be allowed to carry or drag as many items as you wish at one time. It is not expected that you will carry or drag all the items of any task at one time. During this event, time will be added if you rest any of the garbage on the wooden boxes.

For Task 1 you will start in the rest area. At the examiner's signal, the clock will be started. You will carry or drag eight items of garbage (8, 8, 8, 15, 15, 15, 25, and 55 pounds for a total of 149 pounds) through an opening between two wooden boxes to a truck, step over the line on the floor with at least one foot, and place the garbage into the truck. Do not rest the garbage on the wooden boxes.

You will immediately proceed to Task 2. Here you will carry or drag eight items of garbage (8, 8, 15, 15, 15, 15, 25, and 45 pounds for a total of 146 pounds) over an obstacle simulating a snow bank, step over the line on the floor with at least one foot and place the garbage in the truck.

You will immediately proceed to Task 3. Task 3 requires that you carry or drag eight items of garbage (8, 8, 8, 15, 25, 25, 25, and 35 pounds for a total of 149 pounds) around a wooden box, step over the line on the floor with at least one foot, and place the garbage in the truck.

You will immediately proceed to Task 4. You will carry or drag six items of garbage (5, 15, 15, 15, 25, and 65 pounds for a total of 140 pounds) between two wooden boxes, step over the line on the floor with at least one foot, and place the garbage in the truck.

You will immediately proceed to Task 5 where you will carry or drag seven items of garbage (8, 15, 15, 15, 25, 35, and 35 pounds for a total of 148 pounds) over a railroad tie, step over the line on the floor with at least one foot, and place the garbage in the truck. The examiner will stop the clock for Event IV when you have carried or dragged the last piece of garbage over the railroad tie and placed it in the truck. The examiner will then direct you to the rest area where you will rest for another required two minutes before proceeding to Event V.

Event V: Carry or Drag Garbage to a Truck

Event V consists of six tasks which must be performed consecutively. You will be allowed to carry or drag as many items as you wish at one time. It is not expected that you will carry or drag all the items of any one task at the same time.

For Task 1 you will start at the rest area. You will carry or drag eleven items of garbage totalling 143 pounds to a truck, step over the line on the floor with at least one foot, and place the garbage in the truck.

You will immediately proceed to Tasks 2 through 5, which will be performed in the same manner as Task 1 but with differing weights. Task 2 entails six items and 140 pounds; Task 3 entails seven items and 168 pounds; Task 4 entails seven items and 141 pounds; Task 5 entails six items and 143 pounds. In all, the first five tasks of Event V entail moving 37 items at a total weight of 735 pounds.

After finishing the first five tasks, you will immediately proceed to Task 6.

Event VI: Drag Mattress Across Line

For this last task you will drag a full-size mattress without handles eight feet across the line on the floor. The examiner will stop the clock for Event V when the mattress is completely across the line.

This is the end of the New York City physical performance test. Your gloves will be collected and you will be permitted to rest before leaving.

Second Physical Performance Test

While all Sanitation Worker physical fitness tests are designed to measure physical strength and stamina, the procedures used may be very different from one department to the next. The following physical fitness test is very different from the New York City exam just described, but you must be equally fit to pass it.

A total score of 20 is required for passing this test; the scores attained on the five individual tests are added together to obtain your final score.

Event I: Trunk Flexion Test—3 Chances

Candidates will assume a sitting position on the floor with the legs extended at right angles to a line drawn on the floor. The heels should touch the near edge of the lines and be 5 inches apart. The candidate should slowly reach with both hands as far forward as possible on a yardstick that is placed between the legs with the 15-inch mark resting on the near edge of the heel line.

The score is the most distant point (in inches) reached on the yardstick with fingertips.

Rating	Trunk Flexion (inches)	Points
Excellent	22 and over	6
Good	20–21	5
Average	14–19	4
Fair	12-13	3
Poor	10–11	2
Very poor	9 and under	1

Event II: Hand Grip Strength Test—3 Chances

The candidate places the dynamometer (hand grip tester) at the side and without touching the body with any part of the arm, hand or the dynamometer, should grip the dynamometer as hard as possible in one quick movement. The best of the 3 tries will be recorded.

Rating	Hand Grip in Kg.	Points
Excellent	65 and above	6
Good	57–64	5
Average	44–56	4
Fair	37–43	3
Poor	30–36	2
Very Poor	29 and under	1

Event III: Standing Broad Jump—3 Chances

Candidates will be permitted 3 chances in consecutive order, and the longest distance will be credited.

Candidates will be required to jump from a standing position, both feet together. Distance of jump will be recorded from starting point to back of heels. It is each candidate's responsibility to have a nonskid surface on the soles of his or her sneakers.

Rating	Distance	Points
Excellent	7'10" or better	6
Good	7' to 7'9"	5
Average	6'1" to 6'11"	4
Fair	5'6" to 6'0"	3
Poor	5'0" to 5'5"	2
Very Poor	Less than 5	1

Event IV: One Minute Situp Test

The candidate will start by lying on the back with the knees bent so that the heels are about 18 inches away from the buttocks. An examiner will hold the ankles to give support. The candidate will then perform as many correct situps (elbows alternately touching the opposite knee) as possible within a one-minute period. The candidate should return to the starting position (back to floor) between situps.

Rating	Situps in 1 Minute	Points
Excellent	35	6
Good	30–34	5
Average	20–29	4
Fair	15–19	3
Poor	10–14	2
Very Poor	9 and under	1

Event V: Three Minute Step Test

The candidate will step for 3 minutes on a 12-inch bench at a rate of 24 steps per minute. The time will be maintained by a metronome. Immediately after the 3 minutes of stepping, the subject will sit down and relax without talking. A 60-second heart rate count is taken starting 5 seconds after the completion of stepping.

Rating	Pulse	Points
Excellent	75–84	6
Good	85–94	5
Average	95–119	4
Fair	120–129	3
Poor	130–139	2
Very Poor	140 and over	1

Third Physical Performance Test

Here is still one more physical performance test given in a different city. Each subtest is followed by the scoring grid actually used in scoring this physical performance test. These scoring grids give you a good idea of the level of performance expected.

Official Description

The physical test will consist of three competitive subtests. In order to pass, candidates must receive a score above zero on each of the three competitive subtests and a general rating of at least 70% for the competitive physical test.

Medical evidence to allow participation in the physical test may be required and the Department of Personnel reserves the right to exclude from the physical test any candidate who, upon examination of such evidence, is apparently medically unfit. Candidates will take the physical test at their own risk of injury, although every effort will be made to safeguard them. It is recommended that candidates wear rubber-soled, safety-tipped shoes and work clothes.

Candidates in the competitive physical test shall be required to take the test during the prescribed period and are required to complete it on schedule in the prescribed manner regardless of accident, injury, illness or any other circumstances. Failure to comply with this requirement shall result in disqualification. No retest, in whole or in part, shall be permitted.

PHYSICAL FITNESS STANDARDS: 70 percent General Average Required

Event I: Garbage Can—Garbage Bag Carry

Weight: 50% of score

At the signal "GO" the candidate will, as directed, pick up a garbage can (or garbage bag) from a shelf, carry it approximately 35 feet to the end of the course, and place it on a second shelf; pick up a bag (or can) from the second shelf and carry it back approximately 35 feet and place it on the first shelf. After 6 cans and 6 bags have been moved in this alternating fashion from one shelf to another, another 6 cans and 6 bags will be moved from one position at floor level to another position at floor level under a shelf, also a distance of approximately 35 feet, in an alternating fashion. After 12 cans and 12 bags have been moved in this fashion, the candidate will repeat the process starting with the cans and bags that are on the shelves. If time allows, the candidate will then begin to repeat the process a third time. Each can weighs approximately 60 pounds, and each bag weighs approximately 40 pounds. At each end of the course, there will be a shelf approximately 40 inches above the floor and space on the floor under the shelf. The candidate must carry the full weight of the can or bag when moving it. Only one can or bag at a time may be carried. Candidates will be rated on the number of cans and bags they have moved in five minutes.

Only one trial will be allowed.

No credit will be given for carrying a bag or can that has been allowed to touch the floor while being moved or which has not been placed in the designated location.

Completed Operations	50	49	48	47	46	45	44	43	42	41	40	39	38	37
Score	50	49	48	47	46	45	44	43	42	41	40	39	38	37

Completed Operations	36	35	34	33	32	31	30	29	28	27	26	25	24	23
Score	36	35	34	33	32	31	30	29	28	27	26	25	24	23

Completed Operations	22	21	20	19	18	17	16	15	14	13	12	11	10	9
Score	22	21	20	19	18	17	16	15	14	13	12	11	10	9

Completed Operations	8	7	6	5	4	3	2	1	0
Score	8	7	6	5	4	3	2	1	0

Event II: Agility-Climb

Weight: 30% of score

At the signal "GO" the candidate will pick up a garbage can weighing approximately 40 pounds from behind a fence approximately 30 inches high, carry it in an upright position to a maze of obstacles, and proceed through, lifting the can above obstacles where necessary; proceed to a wall approximately 6 feet high, place the garbage can on the designated spot behind another fence approximately 30 inches high; mount the wall using the hand-foot supports and climb down the other side; pick up the garbage can, proceed to the finish line, and place the garbage can on the designated spot at the finish line behind the fence, having maintained the can in the upright position at all times. A penalty of two seconds will be added for each instance of hitting a fence or any of the obstacles forming the maze with the body or with the garbage can. A penalty of four seconds will be added for each instance of dropping the can either at the designated locations or elsewhere.

The faster of two trials will be rated.

Seconds	12 or less	14	16	18	20	22	24	26	28	30	32	34	35	36	37	38
Score		30	29	28	27	26	25	24	23	22	21	20	19	18	17	16	15

Seconds	39	40	41	42	43	44	45	46	47	48 or over
Score	14	13	12	11	10	8	6	4	2	zero

Event III: Garbage Can Lift

Weight: 20% of score

The candidate must raise a garbage can from the floor, place it in an upright position on a table approximately three feet in height, and immediately return the can to the floor, in an upright position, still under control. Each garbage can must be lifted three times in succession in this manner. Candidate will start with a garbage can weighing approximately 60 pounds and then lift three times successively the 70, 80, 90, and 100-pound garbage can or, until the limit of strength or endurance has been reached. All candidates must start with the 60-pound garbage can. The cans must be lifted in the exact order described without skipping any garbage can. Candidates will be scored on the number of operations completed in 70 seconds (consisting of lifting a can, placing it on the table and returning it to the floor). No credit will be given for an operation if the can is dropped or otherwise not kept under control.

The better of two trials shall be rated.

Completed Operations	15	14	13	12	11	10	9	8	7	6	5	4	3	2	1	0
Score	20	19	18	17	16	15	14	13	12	11	10	8	6	4	2	zero

PREPARING FOR THE PHYSICAL TEST

If you are considering a career as a sanitation worker you probably consider yourself a healthy, physically fit person. Even so, it would be wise to consult your doctor before proceeding. Tell your doctor about the type of work you have in mind, describe the physical demands, and ask for an assessment of your potential to withstand these rigors. If your doctor foresees any potential problems, either in passing the exams or in facing the demands of the job, discuss corrective measures and remedial programs. While speaking with your doctor you might describe the physical fitness test as well. You may be able to pick up special tips to prepare yourself to do well. If your doctor does not have a physical conditioning program to recommend, design your own program.

Defining Fitness

Physical fitness is to the human body what fine tuning is to an engine. It enables us to perform up to our potential. Fitness can be described as a condition that helps us look, feel, and do our best. More specifically, it is:

The ability to perform daily tasks vigorously and alertly, with energy left over for enjoying leisure-time activities and meeting emergency demands. It is the ability to endure, to bear up, to withstand stress, to carry on in circumstances where an unfit person could not continue, and is a major basis for good health and well-being.

Physical fitness involves the performance of the heart and lungs, and the muscles of the body. And, since what we do with our bodies also affects what we can do with our minds, fitness influences to some degree qualities such as mental alertness and emotional stability.

As you undertake your fitness program, it's important to remember that fitness is an individual quality that varies from person to person. It is influenced by age, sex, heredity, personal habits, exercise and eating practices. You can't do anything about the first three factors. However, it is within your power to change and improve the others where needed.

Knowing the Basics

Physical fitness is most easily understood by examining its components, or "parts." There is widespread agreement that these four components are basic:

Cardiorespiratory Endurance—the ability to deliver oxygen and nutrients to tissues, and to remove wastes, over sustained periods of time. Long runs and swims are among the methods employed in measuring this component.

Muscular Strength—the ability of a muscle to exert force for a brief period of time. Upper-body strength, for example, can be measured by various weight-lifting exercises.

Muscular Endurance—the ability of a muscle, or a group of muscles, to sustain repeated contractions or to continue applying force against a fixed object. Pushups are often used to test endurance of arm and shoulder muscles.

Flexibility—the ability to move joints and use muscles through their full range of motion. The sit-and-reach test is a good measure of flexibility of the lower back and backs of the upper legs.

Body Composition—is often considered a component of fitness. It refers to the makeup of the body in terms of lean mass (muscle, bone, vital tissue, and organs) and fat mass. An optimal ratio of fat to lean mass is an indication of fitness, and the right types of exercises will help you decrease body fat and increase or maintain muscle mass.

A Workout Schedule

How often, how long, and how hard you exercise, and what kinds of exercises you do should be determined by what you are trying to accomplish. Your goals, your present fitness level, age, health, skills, interest, and convenience are among the factors you should consider. For example, an athlete training for high-level competition would follow a different program than a person whose goals are good health and the ability to meet work and recreational needs.

Your exercise program should include something from each of the four basic fitness components described previously. Each workout should begin with a warmup and end with a cool-down. As a general rule, space your workouts throughout the week and avoid consecutive days of hard exercise.

Here are the amounts of activity necessary for the average, healthy person to maintain a minimum level of overall fitness. Included are some of the popular exercises for each category.

Warmup—5–10 minutes of exercises such as walking, slow jogging, knee lifts, arm circles or trunk rotations. Low intensity movements that simulate movements to be used in the activity can also be included in the warmup.

Muscular Strength—a minimum of two 20-minute sessions per week that include exercises for all the major muscle groups. Lifting weights is the most effective way to increase strength.

Muscular Endurance—at least three 30-minute sessions each week that include exercises such as calisthenics, pushups, situps, pullups, and weight training for all the major muscle groups.

Cardiorespiratory Endurance—at least three 20-minute bouts of continuous aerobic (activity requiring oxygen) rhythmic exercise each week. Popular aerobic conditioning activities include brisk walking, jogging, swimming, cycling, rope-jumping, rowing, cross-country skiing, and some continuous action games like racquetball and handball.

Flexibility—10–12 minutes of daily stretching exercises performed slowly, without a bouncing motion. This can be included after a warmup or during a cooldown.

Cool Down—a minimum of 5–10 minutes of slow walking, low-level exercise, combined with stretching.

A Matter of Principle

The keys to selecting the right kinds of exercises for developing and maintaining each of the basic components of fitness are found in these principles:

Specificity—pick the right kind of activities to affect each component. Strength training results in specific strength changes. Also, train for the specific activity you're interested in. For example, optimal swimming performance is best achieved when the muscles involved in swimming are trained for the movements required. It does not necessarily follow that a good runner is a good swimmer.

Overload—work hard enough, at levels that are vigorous and long enough to overload your body above its resting level, to bring about improvement.

Regularity—you can't hoard physical fitness. At least three balanced workouts a week are necessary to maintain a desirable level of fitness.

Progression—increase the intensity, frequency and/or duration of activity over periods of time in order to improve.

Some activities can be used to fulfill more than one of your basic exercise requirements. For example, in addition to increasing cardiorespiratory endurance, running builds muscular endurance in the legs, and swimming develops the arm, shoulder, and chest muscles. If you select the proper activities, it is possible to fit parts of your muscular endurance workout into your cardiorespiratory workout and save time.

Measuring Your Heart Rate

Heart rate is widely accepted as a good method for measuring intensity during running, swimming, cycling, and other aerobic activities. Exercise that doesn't raise your heart rate to a certain level and keep it there for 20 minutes won't contribute significantly to cardiovascular fitness.

The heart rate you should maintain is called your **target heart rate**. There are several ways of arriving at this figure. One of the simplest is: **maximum heart rate** (220 − age) × 70%. Thus, the target heart rate for a 40-year-old would be 126.

Some methods for figuring the target rate take individual differences into consideration. Here is one of them:

1. Subtract age from 220 to find **maximum heart rate**.
2. Subtract resting heart rate (see below) from maximum heart rate to determine **heart rate reserve**.
3. Take 70% of heart rate reserve to determine **heart rate raise**.
4. Add heart rate raise to resting heart rate to find **target rate**.

Resting heart rate should be determined by taking your pulse after sitting quietly for five minutes. When checking heart rate during a workout, take your pulse within five seconds after interrupting exercise because it starts to go down once you stop moving. Count pulse for 10 seconds and multiply by six to get the per-minute rate.

Physical fitness is not achieved in a day, a week, or even a month. Good physical fitness is built up over a lifetime of healthful habits, good diet, and adequate exercise. However, it is never too late to begin physical conditioning. There is a considerable time lag between filing to take the sanitation worker exam and the date of the physical performance test. The city of New York makes the following suggestions to applicants.

CONDITIONING TECHNIQUES

The physical performance test associated with this job is fairly demanding. It involves a series of job-like events which include such movements as climbing up and down ladders, and lifting and carrying objects. Depending on your current fitness level, these events may range from fairly easy to very difficult. There are some basic things you should do in preparation to take the physical test.

FLEXIBILITY

Many of the events involve climbing, reaching, stepping over and around objects, etc. The best way of improving flexibility is by doing some regular basic stretching exercises. To get the most benefit from stretching and to avoid injury, stretching should be done without bouncing. In other words, stretch to the point where you feel slight tension on the muscles and hold it there for 10-20 seconds without rocking or bouncing. This should be repeated 3 to 5 times for each muscle group. During the stretching you should breathe easily and feel relaxed. While any stretching will help, the following exercises are especially recommended:

A. NECK   Tilt the head to the left, right, and front. Be sure to bring the head to the upright position between each movement.

Avoid tilting the head back or doing full neck circles.

B. BACK   (1) While lying on your back with your knees bent and feet flat on the ground, hug one knee to the chest while keeping the other foot flat on the ground. Return the first foot to the ground and hug second knee to chest. Then hug both knees to the chest.
(2) While sitting on the ground with your legs straight out, reach for your toes.

Avoid rigidly locking the knees when doing exercise (2).

C. THIGHS   (1) While standing, face a wall and place one hand on the wall for balance. With the other hand, grasp and pull up on your ankle to stretch your quadriceps. (2) To stretch the hamstrings, put one foot on a table, bench, or chair, lean forward, slowly sliding your hands toward the ankle. For these exercises hold at the point of tension for 10-20 seconds as described in II above.

Avoid any stretches which bend the knee at an awkward angle.

MUSCULAR ENDURANCE

Muscular endurance is very different from muscle strength. Strength involves lifting extremely heavy objects only one or two times and is a measure of the maximum weight you can lift. Muscular endurance involves frequent lifting of lighter loads. Endurance, especially for the upper body and arm muscles and abdominal (stomach) muscles, is better related to successful performance on the test. The following simple exercises are recommended:

A. Pushups. Build up to 20–30 per day.
B. Curlups. With the knees bent, do the first part of a situp but stop at the point where more muscles would be involved in going all the way up. Build up to 30–40 per day.
C. Chinups and pullups. If available. Sets of 10 reversing the hand position between sets.
D. Arm circles and side lateral raises with weights. Holding soup cans or something slightly heavier, do small arm circles with arms horizontally out at the side. Repeat with arms horizontally in front of the body. Finally, repeated lifts (10–50) from arms at the sides to shoulder level.

GENERAL HEALTH HABITS

Stop smoking and excessive intake of alcohol. Moderate dieting is encouraged if overweight, but avoid starvation diets. Get plenty of sleep the night prior to a test. If you do no aerobic activity, beginning a program of regular walking (30–60 minutes per day) may be beneficial.

If self-discipline with daily exercise is not your strong point, you may need a regimented daily program. You might find the following suggestions prepared by the President's Council on Physical Fitness helpful in planning a fitness program for your own needs and time requirements, or you might even follow the program just as printed.

THE PROGRAM

Getting Set—Orientation Workouts

With the series of mild exercises listed in the charts on pages 146 and 158, you can get yourself ready—without severe aches or pains—for the progressive conditioning program.

Plan to spend a minimum of one week for preliminary conditioning. Don't hesitate to spend two weeks or three if necessary for you to limber up enough to accomplish all the exercises easily and without undue fatigue.

The sanitation worker physical performance test is identical for both men and women because all sanitation workers must be able to perform the same tasks. There are far fewer women sanitation workers than men precisely because many women do not have the physical strength for the work. The women who can meet the physical standards share equally in duties, responsibilities, risks, and hard work.

There are, of course, physiological differences between men and women. Some conditioning exercises in this program are modified in recognition of these differences. Women with the potential to pass the sanitation worker physical performance test should find that the women's program described here, if followed faithfully, will prepare them well. The men's program, which is similar, begins on page 155.

The Program For Women

This program assumes that you have not been putting all of your muscles to consistent use and that you are starting from close to "couch potato" status. If you are already in pretty good shape, you might be able to to start more quickly. But do not overdo. A gradual build-up makes sense.

The program starts with an orientation or "get-set" series of exercises that will allow you to bring all major muscles into use easily and painlessly.

There are then five graded levels.

As you move from one to the next, you will be building toward a practical and satisfying level of fitness.

By building gradually, progressively, you will be building soundly.

What the Exercises Are For

There are three general types—warmup exercises, conditioning exercises, and circulatory activities.

The warmup exercises stretch and limber up the muscles and speed up the action of the heart and lungs, thus preparing the body for greater exertion and reducing the possibility of unnecessary strain.

The conditioning exercises are systematically planned to tone up abdominal, back, leg, arm, and other major muscles.

The circulatory activities produce contractions of large muscle groups for relatively longer periods than the conditioning exercises to stimulate and strengthen the circulatory and respiratory systems.

The plan calls for 10 mild exercises during the orientation period and, thereafter, the warmup exercises and the seven conditioning exercises listed for each level. The first six exercises of the orientation program are used as warmup exercises throughout the graded levels.

When it comes to the circulatory activities, you choose one each workout. Alternately running and walking . . . skipping rope . . . running in place. All are effective. You can choose running and walking on a pleasant day and one of the others for use indoors when the weather is inclement. You can switch about for variety.

How You Progress

A sound physical conditioning program should take into account your individual tolerance—your ability to execute a series of activities without undue discomfort or fatigue. It should provide for developing your tolerance by increasing the work load so you gradually become able to achieve more and more with less and less fatigue and with increasingly rapid recovery.

As you move from level to level, some exercises will be modified so they call for increased effort.

Others will remain the same but you will build more strength and stamina by increasing the number of repetitions.

You will be increasing your fitness another way as well.

At level 1, your objective will be to gradually reduce, from workout to workout, the "breathing spells" between exercises until you can do the seven conditioning exercises without resting. You will proceed in the same fashion with the more difficult exercises and increased repetitions at succeeding levels.

You will find the program designed—the progression carefully planned—to make this feasible. You will be able to proceed at your own pace, competing with yourself rather than with anyone else—and this is of great importance for sound conditioning.

Note: Gradually speeding up, from workout to workout, the rate at which you do each exercise will provide greater stimulation for the circulatory and respiratory systems and also help to keep your workouts short. However, the seven conditioning exercises should not be a race against time. Perform each exercise correctly to insure maximum benefit.

How Long at Each Level

Your objective at each level will be to reach the point where you can do all the exercises called for, for the number of times indicated, without resting between exercises.

But, start slowly.

It cannot be emphasized enough that by moving forward gradually you will be moving forward solidly, avoiding sudden strains and excesses that could make you ache and hold you back for several days.

If you find yourself at first unable to complete any exercises—to do continuously all the repetitions called for—stop when you encounter difficulty. Rest briefly, then take up where you left off and complete the count. If you have difficulty at first, there will be less and less with succeeding workouts.

Stay at each level for at least three weeks. If you have not passed the prove-out test at the end of that time, continue at the same level until you do. The prove-out test calls for performing—in three consecutive workouts—the seven conditioning exercises without resting and satisfactorily fulfilling the requirement for one circulatory activity.

Measuring Your Progress

You will, of course, be able to observe the increase in your strength and stamina from week to week in many ways—including the increasing facility with which you do the exercises at a given level.

In addition, there is a 2-minute step test you can use to measure and keep a running record of the improvement in your circulatory efficiency, one of the most important of all aspects of fitness.

The immediate response of the cardiovascular system to exercise differs markedly between well-conditioned individuals and others. The test measures the response in terms of pulse rate taken shortly after a series of steps up and down onto a bench or chair.

Although it does not take long, it is necessarily vigorous. Stop if you become overly fatigued while taking it. You should not try it until you have completed the orientation period.

The Step Test

Use any sturdy bench or chair 15–17 inches in height.

> Count 1—Place right foot on bench.
> Count 2—Bring left foot alongside of right and stand erect.
> Count 3—Lower right foot to floor.
> Count 4—Lower left foot to floor.
>
> REPEAT the 4-count movement 30 times a minute for two minutes.
>
> THEN sit down on bench or chair for two minutes.
>
> FOLLOWING the 2-minute rest, take your pulse for 30 seconds. Double the count to get the per-minute rate. (You can find the pulse by applying middle and index finger of one hand firmly to the inside of the wrist of the other hand, on the thumb side.)

Record your score for future comparisons. In succeeding tests—about once every two weeks—you probably will find your pulse rate becoming lower as your physical condition improves.

Three important points:

1. For best results, do not engage in physical activity for at least 10 minutes before taking the test. Take it at about the same time of day and always use the same bench or chair.
2. Remember that pulse rates vary among individuals. This is an individual test. What is important is not a comparison of your pulse rate with that of anybody else—but rather a record of how your own rate is reduced as your fitness increases.
3. As you progress, the rate at which your pulse is lowered should gradually level off. This is an indication that you are approaching peak fitness.

Your Progress Records

Charts are provided for the orientation program and for each of the five levels.

They list the exercises to be done and the goal for each exercise in terms of number of repetitions, distance, etc.

They also provide space in which to record your progress—(1) in completing the recommended 15 workouts at each level, (2) in accomplishing the three prove-out workouts before moving on to a succeeding level, and (3) in the results as you take the step test from time to time.

A sample chart and progress record for one of the five levels is shown on the next page.

You do the warmup exercises and the conditioning exercises along with one circulatory activity for each workout.

Check off each workout as you complete it. The last three numbers are for the prove-out workouts, in which the seven conditioning exercises should be done without resting. Check them off as you accomplish them.

You are now ready to proceed to the next level.

As you take the step test—at about 2-week intervals—enter your pulse rate.

When you move on to the next level, transfer the last pulse rate from the preceding level. Enter it in the margin to the left of the new progress record and circle it so it will be convenient for continuing reference.

SAMPLE	GOAL
Warmup Exercises	**Exercises 1–6 of Orientation program**
Conditioning Exercises	**Uninterrupted repetitions**
1. Bend and stretch	10
2. Sprinter	6
3. Sitting stretch	15
4. Knee pushup	12
5. Sit-up (fingers laced)	10
6. Leg raiser	10 each leg
7. Flutter kick	30
Circulatory activity (choose one each workout)	
Jog-walk (jog 50, walk 50)	½ mile
Rope (skip 30 secs.; rest 60 secs.)	3 series
Run in place (run 100, hop 25–2 cycles)	3 minutes
Water activities (see page 169)	

Your progress record	1 2 3 4 5 6 7 8 9 10 11 12	13 14 15
Step test (pulse)		Prove-out workouts

ORIENTATION PROGRAM: WOMEN	GOAL
Conditioning Exercises	**Repetitions**
*1. Bend and stretch	10
*2. Knee lift	10 left, 10 right
*3. Wing stretcher	20
*4. Half knee bend	10
*5. Arm circles	15 each way
*6. Body bender	10 left, 10 right
7. Prone arch	10
8. Knee pushup	6
9. Head and shoulder curl	5
10. Ankle stretch	15
Circulatory activity (choose one each workout)	
Walking	½ mile
Rope (skip 15 sec.; rest 60 sec.)	3 series

*The first six exercises of the orientation program will be used as warmup exercises throughout the graded levels.

Step Test Record—After completing the orientation program, take the 2-minute step test. Record your pulse rate here: _____ . This will be the base rate with which you can make comparisons in the future.

1. Bend and Stretch

Starting position: Stand erect, feet shoulder-width apart.

Action: Count 1. Bend trunk forward and down, flexing knees. Stretch gently in attempt to touch fingers to toes or floor. Count 2. Return to starting position.

Note: Do slowly, stretch and relax at intervals rather than in rhythm.

2. Knee Lift

Starting position: Stand erect, feet together, arms at sides.

Action: Count 1. Raise left knee as high as possible, grasping leg with hands and pulling knee against body while keeping back straight. Count 2. Lower to starting position. Counts 3 and 4. Repeat with right knee.

3. Wing Stretcher

Starting position: Stand erect, elbows at shoulder height, fists clenched in front of chest.
Action: Count 1. Thrust elbows backward vigorously without arching back. Keep head erect, elbows at shoulder height. Count 2. Return to starting position.

4. Half Knee Bend

Starting position: Stand erect, hands on hips.
Action: Count 1. Bend knees halfway while extending arms forward, palms down. Count 2. Return to starting position.

5. Arm Circles

Starting position: Stand erect, arms extended sideward at shoulder height, palms up.
Action: Describe small circles backward with hands. Keep head erect. Do 15 backward circles. Reverse, turn palms down and do 15 small circles forward.

6. Body Bender

Starting position: Stand, feet shoulder-width apart, hands behind neck, fingers interlaced.
Action: Count 1. Bend trunk sideward to left as far as possible, keeping hands behind neck. Count 2. Return to starting position. counts 3 and 4. Repeat to the right.

7. Prone Arch

Starting position: Lie face down, hands tucked under thighs.

Action: Count 1. Raise head, shoulders, and legs from floor. Count 2. Return to starting position.

8. Knee Pushup

Starting position: Lie on floor, face down, legs together, knees bend with feet raised off floor, hands on floor under shoulders, palms down.
Action: Count 1. Push upper body off floor until arms are fully extended and body is in straight line from head to knees. Count 2. Return to starting position.

9. Head and Shoulder Curl

Starting position: Lie on back, hands tucked under small of back, palms down.
Action: Count 1. Tighten abdominal muscles, lift head and pull shoulders and elbows off floor. Hold for four seconds. Count 2. Return to starting position.

10. Ankle Stretch

Starting position: Stand on a stair, large book, or block of wood, with weight on balls of feet and heels raised.
Action: Count 1. Lower heels. Count 2. Raise heels.

Circulatory Activities

WALKING—Step off at a lively pace, swing arms and breathe deeply. *ROPE*—Any form of skipping or jumping is acceptable. Gradually increase the tempo as your skill and condition improve.

LEVEL ONE: WOMEN		GOAL
Warmup Exercises		**Exercises 1–6 of Orientation Program**
Conditioning Exercises		**Uninterrupted repetitions**
1. Toe touch		5
2. Sprinter		8
3. Sitting stretch		10
4. Knee pushup		8
5. Situp (arms extended)		5
6. Leg raiser		5 each leg
7. Flutter kick		20
Circulatory activity (choose one each workout)		
Walking (120 steps a minute)		½ mile
Rope (skip 30 secs.; rest 60 secs.)		2 series
Run in place (run 50, straddle hop 10–2 cycles)		2 minutes
Water activities (see page 169)		
Your progress record	1 2 3 4 5 6 7 8 9 10 11 12	13 14 15
Step test (pulse)		Prove-out workouts

1. Toe Touch

Starting Position: Stand at attention.
Action: Count 1. Bend trunk forward and down, keeping knees straight, touching fingers to ankles. Count 2. Bounce and touch fingers to top of feet. Count 3. Bounce and touch fingers to toes. Count 4. Return to starting position.

2. Sprinter

Starting position: Squat, hands on floor, fingers pointed forward, left leg fully extended to rear.
Action: Count 1. Reverse position of feet in bouncing movement, bringing left foot to hands, extending right leg backward—all in one motion. Count 2. Reverse feet again, returning to starting position.

3. Sitting Stretch

Starting position: Sit, legs spread apart, hands on knees.
Action: Count 1. Bend forward at waist, extending arms as far forward as possible. Count 2. Return to starting position.

4. Knee Pushup

Starting position: Lie on floor, face down, legs together, knees bent with feet raised off floor, hands on floor under shoulders, palms down.
Action: Count 1. Push upper body off floor until arms are fully extended and body is in straight line from head to knees. Count 2. Return to starting position.

5. Situp (Arms Extended)

Starting position: Lie on back, legs straight and together, arms extended beyond head.
Action: Count 1. Bring arms forward over head, roll up to sitting position, sliding hands along legs, grasping ankles. Count 2. Roll back to starting position.

6. Leg Raiser

Starting position: Right side of body on floor, head resting on right arm.
Action: Lift left leg about 24″ off floor, then lower it. Do required number of repetitions. Repeat on other side.

7. Flutter Kick

Starting position: Lie face down, hands tucked under thighs.
Action: Arch the back, bringing chest and head up, then flutter kick continuously, moving the legs 8–10″ apart. Kick from hips and with knees slightly bent. Count each kick as one.

Circulatory Activities

WALKING—Maintain a pace of 120 steps per minute for a distance of ½ mile. Swing arms and breathe deeply.
ROPE—Skip or jump rope continuously using any form for 30 seconds and then rest 60 seconds. Repeat 2 times.

RUN IN PLACE—Raise each foot at least 4″ off the floor and jog in place. Count 1 each time left foot touches floor. Complete number of running steps called for in chart, then do specified number of straddle hops. Complete 2 cycles of alternate running and hopping for time specified on chart.
STRADDLE HOP—Starting position: At attention.
Action: Count 1. Swing arms sideward and upward, touching hands above head (arms straight) while simultaneously moving feet sideward and apart in a single jumping motion. Count 2. Spring back to starting position. Two counts in one hop.

LEVEL TWO: WOMEN — GOAL

Warmup Exercises	Exercises 1–6 of Orientation Program
Conditioning Exercises	**Uninterrupted repetitions**
1. Toe touch	10
2. Sprinter	12
3. Sitting stretch	15
4. Knee pushup	12
5. Situp (fingers laced)	10
6. Leg raiser	10 each leg
7. Flutter kick	30

Circulatory activity (choose one each workout)

Jog-walk (jog 50, walk 50)	½ mile
Rope (skip 30 secs.; rest 60 secs.)	3 series
Run in place (run 80, hop 15–2 cycles)	3 minutes

Water activities (see page 169)

Your progress record	1 2 3 4 5 6 7 8 9 10 11 12	13 14 15
Step test (pulse)		Prove-out workouts

1. Toe Touch

Starting position: Stand at attention.
Action: Count 1. Bend trunk forward and down, keeping knees straight, touching fingers to ankles. Count 2. Bounce and touch fingers to top of feet. Count 3. Bounce and touch fingers to toes. Count 4. Return to starting position.

2. Sprinter

Starting position: Squat, hands on floor, fingers pointed forward, left leg fully extended to rear.
Action: Count 1. Reverse position of feet in bouncing movement, bringing left foot to hands, extending right leg backward—all in one motion. Count 2. Reverse feet again, returning to starting position.

3. Sitting Stretch

Starting position: Sit, legs spread apart, hands on knees.

Action: Count 1. Bend forward at waist, extending arms as far forward as possible. Count 2. Return to starting position.

4. Knee Pushup

Starting position: Lie on floor, face down, legs together, knees bent with feet raised off floor, hands on floor under shoulders, palms down.

Action: Count 1. Push upper body off floor until arms are fully extended and body is in straight line from head to knees. Count 2. Return to starting position.

5. Situp (Fingers Laced)

Starting position: Lie on back, legs straight and feet spread approximately 1' apart. Fingers laced behind neck.

Action: Count 1. Curl up to sitting position and turn trunk to left. Touch right elbow to left knee. Count 2. Return to starting position. Count 3. Curl up to sitting position and turn trunk to right. Touch left elbow to right knee. Count 4. Return to starting position. Score one situp each time you return to starting position. Knees may be bent as necessary.

6. Leg Raiser

Starting position: Right side of body on floor, head resting on right arm.

Action: Lift left leg about 25″ off floor, then lower it. Do required number of repetitions. Repeat on other side.

7. Flutter Kick

Starting position: Lie face down, hands tucked under thighs.

Action: Arch the back, bringing chest and head up, then flutter kick continuously, moving the legs 8—10″ apart. Kick from hips with knees slightly bent. Count each kick as one.

Circulatory Activities

JOG-WALK—Jog and walk alternately for number of paces indicated on chart for distance specified.

ROPE—Skip or jump rope continously using any form for 30 seconds and then rest 60 seconds. Repeat 3 times.

RUN IN PLACE—Raise each foot at least 4″ off floor and jog in place. Count 1 each time left foot touches floor. Complete number of running steps called for in chart, then do specified number of straddle hops. Complete 2 cycles of alternate running and hopping for time specified on chart.

STRADDLE HOP—Starting position: At attention.

Action: Count 1. Swing arms sideward and upward, touching hands above head (arms straight) while simultaneously moving feet sideward and apart in a single jumping motion. Count 2. Spring back to starting position. Two counts in one hop.

LEVEL THREE: WOMEN   **GOAL**

Warmup	Exercises 1–6 of Orientation Program
Conditioning Exercises	**Uninterrupted repetitions**
1. Toe touch	20
2. Sprinter	16
3. Sitting stretch (fingers laced)	15
4. Knee pushup	20
5. Situp (arms extended, knees up)	15
6. Leg raiser	16 each leg
7. Flutter kick	40

Circulatory activity (choose one each workout)

Jog-walk (jog 50, walk 50)	¾ mile
Rope (skip 45 secs.; rest 30 secs.)	3 series
Run in place (run 110, hop 20–2 cycles)	4 minutes

Water activities (see page 169)

Your progress record	1 2 3 4 5 6 7 8 9 10 11 12	13 14 15
Step test (pulse)		Prove-out workouts

1. Toe Touch

Starting position: Stand at attention.
Action: 1.Bend trunk forward and down, keeping knees straight, touching fingers to ankles. 2. Bounce and touch fingers to top of feet. Count 3. Bounce and touch fingers to toes. Count 4. Return to starting position.

2. Sprinter

Starting position: Squat, hands on floor, fingers pointed forward, left leg fully extended to rear.
Action: Count 1. Reverse position of feet in bouncing movement, bringing left foot to hands, extending right leg backward all in one motion. Count 2. Reverse feet again, returning to starting position.

3. Sitting Stretch
(Fingers Laced)

Starting position: Sit, legs spread apart, fingers laced behind neck.
Action: Count 1. Bend forward at waist, reaching elbows as close to floor as possible. Count 2. Return to starting position.

4. Knee Pushup

Starting position: Lie on floor, face down, legs together, knees bent with feet raised off floor, hands on floor under shoulders, palms down.
Action: Count 1. Push upper body off floor until arms are fully flexed and body is in straight line from head to knees. Count 2. Return to starting position.

5. Situp
(Arms Extended, Knees Up)

Starting position: Lie on back, legs straight, arms extended overhead.
Action: Count 1. Sit up, reaching forward with arms encircling knees while pulling them tightly to chest. Count 2. Return to starting position. Do this exercise rhythmically, without breaks in the movement.

6. Leg Raiser

Starting position: Right side of body on floor, head resting on right arm.
Action: Lift left leg about 24″ off floor, then lower it. Do required number of repetitions. Repeat on other side.

7. Flutter Kick

Starting position: Lie face down, hands tucked under thighs.
Action: Arch the back, bringing chest and head up. Then flutter kick continuously, moving the legs 8–10″ apart. Kick from hips with knees slightly bent. Count each kick as one.

Circulatory Activities

JOG-WALK—Jog and walk alternately for number of paces indicated on chart for distance specified.
ROPE—Skip or jump rope continuously using any form for 45 seconds and then rest 30 seconds. Repeat 3 times.

RUN IN PLACE—Raise each foot at least 4″ off floor and jog in place. Count 1 each time left foot touches floor. Complete number of running steps called for in chart, then do specified number of straddle hops. Complete 2 cycles of alternate running and hopping for time specified on chart.
STRADDLE HOP—*Starting position:* At attention.
Action: Count 1. Swing arms sideward and upward, touching hands above head (arms straight) while simultaneously moving feet sideward and apart in a single jumping motion. Count 2. Spring back to starting position. Two counts in one hop.

LEVEL FOUR: WOMEN   GOAL

Warmup Exercises	Exercises 1–6 of Orientation Program
Conditioning Exercises	**Uninterrupted repetitions**
1. Toe touch (twist and bend)	15 each side
2. Sprinter	20
3. Sitting stretch (alternate)	20
4. Pushup	8
5. Situp (arms crossed, knees bent)	20
6. Leg raiser (whip)	10 each leg
7. Prone arch (arms extended)	15
Circulatory activity (choose one each workout)	
Jog-walk (jog 100; walk 50)	1 mile
Rope (skip 60 secs.; rest 30 secs.)	3 series
Run in place (run 145, hop 25–2 cycles)	5 minutes
Water activities (see page 169)	

Your progress record	1 2 3 4 5 6 7 8 9 10 11 12	13 14 15
Step test (pulse)		Prove-out workouts

1. Toe Touch (Twist and Bend)

Starting position: Stand, feet shoulder-width apart, arms extended over head, thumbs interlocked.
Action: Count 1. Twist trunk to right and touch floor inside right foot with fingers of both hands. Count 2. Touch floor outside toes of right foot. Count 3. Touch floor outside heel of right foot. Count 4. Return to starting position, sweeping trunk and arms upward in a wide arc. On the next four counts, repeat action to left side.

2. Sprinter

Starting position: Squat, hands on floor, fingers pointed forward, left leg fully extended to rear.
Action: Count 1. Reverse position of feet in bouncing movement, bringing left foot to

hands, extending right leg backward—all in one motion. Count 2. Reverse feet again, returning to starting position.

3. Sitting Stretch (Alternate)

Starting position: Sit, legs spread apart, fingers laced behind neck, elbows back.
Action: Count 1. Bend forehead to left, touching forward to left knee. Count 2. Return to starting position. Counts 3 and 4. Repeat to right. Score one repetition each time you return to starting position. Knees may be bent if necessary.

4. Pushup

Starting position: Lie on floor, face down, legs together, hands on floor under shoulders with fingers pointing straight ahead.
Action: Count 1. Push body off floor by extending arms so that weight rests on hands and toes. Count 2. Lower the body until chest touches floor.
Note: Body should be kept straight, buttocks should not be raised, abdomen should not sag.

5. Situp
(Arms Crossed, Knees Bent)

Starting position: Lie on back, arms crossed on chest, hands grasping opposite shoulders, knees bent to right angle, feet flat on floor.
Action: Count 1. Curl up to sitting position. Count 2. Return to starting position.

6. Leg Raiser (Whip)

Starting position: Right side of body on floor, right arm supporting head.

Action: Whip left leg up and down rapidly lifting as high as possible off the floor. Count each whip as one. Reverse position and whip right leg up and down.

7. Prone Arch
(Arms Extended)

Starting position: Lie face down, legs straight and together, arms extended to sides at shoulder level.
Action: Count 1. Arch the back, bringing arms, chest and head up, and raising legs as high as possible. Count 2. Return to starting position.

Circulatory Activities

JOG-WALK—Jog and walk alternately for number of paces indicated on chart for distance specified.
ROPE—Skip or jump rope continuously using any form for 60 seconds and then rest 30 seconds. Repeat 3 times.
RUN IN PLACE—Raise each foot at least 4" off floor and jog in place. Count 1 each time left foot touches floor. Complete number of running steps called for in chart, then do specified number of straddle hops. Complete 2 cycles of alternate running and hopping for time specified on chart.
STRADDLE HOP—*Starting position:* At attention.
Action: Count 1. Swing arms sideward and upward, touching hands above head (arms straight) while simultaneously moving feet sideward and apart in a single jumping motion. Count 2. Spring back to starting position. Two counts in one hop.

LEVEL FIVE: WOMEN	GOAL
Warmup Exercises	**Exercises 1–6 of Orientation Program**
Conditioning Exercises	**Uninterrupted repetitions**
1. Toe touch (twist and bend)	25 each side
2. Sprinter	24
3. Sitting stretch (alternate)	26
4. Pushup	15
5. Situp (fingers laced, knees bent)	25
6. Leg raiser (on extended arm)	10 each side
7. Prone arch (fingers laced)	25
Circulatory activity (choose one each workout)	
Jog-run	1 mile
Rope (skip 2 mins.; rest 45 secs.)	2 series
Run in place (run 180, hop 30–2 cycles)	6 minutes
Water activities (see page 169)	
Your·progress record 1 2 3 4 5 6 7 8 9 10 11 12	13 14 15
	Prove-out
Step test (pulse)	workouts

1. Toe Touch (Twist and Bend)

Starting position: Stand, feet shoulder-width apart, arms extended over head, thumbs interlocked.

Action: Count 1. Twist trunk to right and touch floor inside right foot with fingers of both hands. Count 2. Touch floor outside toes of right foot. Count 3. Touch floor outside heel of right foot. Count 4. Return to starting position, sweeping trunk and arms upward in a wide arc. On the next four counts, repeat action to left side.

2. Sprinter

Starting position: Squat, hands on floor, fingers pointed forward, left leg fully extended to rear.

Action: Count 1. Reverse position of feet in bouncing movement, bringing left foot to hands, extending right leg backward—all in one motion. Count 2. Reverse feet again, returning to starting position.

3. Sitting Stretch (Alternate)

Starting position: Sit, legs spread apart, fingers behind neck, elbows back.

Action: Count 1. Bend forward to left, touching forehead to left knee. Count 2. Return to starting position. Counts 3 and 4. Repeat to right. Score one repetition each time you return to starting position. Knees may be bent if necessary.

4. Pushup

Starting position: Lie on floor, face down, legs together, hands on floor under shoulders with fingers pointing straight ahead.

Action: Count 1. Push body off floor by extending arms so that weight rests on hands and toes. Count 2. Lower the body until chest touches floor.

Note: Body should be kept straight, buttocks should not be raised, abdomen should not sag.

5. Situp (Fingers Laced, Knees Bent)

Starting position: Lie on back, fingers laced behind neck, knees bent, feet flat on floor.

Action: Count 1. Sit up, turn trunk to right, touch left elbow to right knee. Count 2. Return to starting position. Count 3. Sit up, turn trunk to left, touch right elbow to left knee. Count 4. Return to starting position.

Score one each time you return to starting position.

6. Leg Raiser
(On Extended Arm)

Starting position: Body rigidly supported by extended right arm and foot. Left arm is held behind head.
Action: Count 1. Raise left leg high. Count 2. Return to starting position slowly. Repeat on other side. Do required number of repetitions.

7. Prone Arch
(Fingers Laced)

Starting position: Lie face down, fingers laced behind neck.
Action: Count 1. Arch the back, legs and chest off floor. Count 2. Extend arms fully forward. Count 3. Return hands to behind neck. Count 4. Flatten body to floor.

(End of program for women, see pages 167–170 for more instruction.)

Circulatory Activities

JOG-RUN—Jog and run alternately for distance specified on chart.
ROPE—Skip or jump rope continuously using any form for 2 minutes and then rest 45 seconds. Repeat 2 times.
RUN IN PLACE—Raise each foot at least 4″ off floor and jog in place. Count 1 each time left foot touches floor. Complete number of running steps called for in chart, then do specified number of straddle hops. Complete 2 cycles of alternate running and hopping in time specified on the chart.
STRADDLE HOP—Starting position: At attention.
Action: Count 1. Swing arms sideward and upward, touching hands above head (arms straight) while simultaneously moving feet sideward and apart in a single jumping motion. Count 2. Spring back to starting position. Two counts in one hop.

The Program For Men

This program assumes you have not—recently and consistently—been exposed to vigorous, all-around physical activity—which could be true even if you play golf once or twice a week or engage in some other sport. No one sport provides for balanced development of all parts of the body.

The plan starts with an orientation—"get-set"—series of mild exercises to limber up all major muscle groups and help assure a painless transition.

There are then five graded levels.

As you move up from one to the next, you will be building toward a practical and satisfactory level of fitness.

By building gradually and progressively, you will be building soundly.

What the Exercises Are For

There are three general types—warmup exercises, conditioning exercises, and circulatory activities.

The warmup exercises stretch and limber up the muscles and speed up the action of the heart and lungs, thus preparing the body for greater exertion and reducing the possibility of unnecessary strain.

The conditioning exercises are systematically planned to tone up abdominal, back, leg, arm, and other major muscles.

The circulatory activities produce contractions of large muscle groups for relatively longer periods than the conditioning exercises—to stimulate and strengthen the circulatory and respiratory systems.

The plan calls for 10 mild exercises during the orientation period and, thereafter, the warmup exercises and the seven conditioning exercises listed for each level. The first six exercises of the orientation program are used as warmup exercises throughout the graded levels.

When it comes to the circulatory activities, you select one each workout. Alternately running and walking . . . skipping rope . . . running in place. All are effective. You can switch about for variety.

How You Progress

Right now, you have limited tolerance for exercise; you can do just so much without discomfort and fatigue.

A sound conditioning program should gradually stretch your tolerance. It should give unused or little-used muscles moderate tasks at first, then make the tasks increasingly more demanding so you become able to achieve more and more with less and less fatigue and with increasingly rapid recovery.

As you move from level to level, some exercises will be modified so they call for more effort. Others will remain the same but you will build strength and stamina by increasing the number of repetitions.

You will be increasing your fitness another way as well.

At level 1, your objective will be to gradually reduce, from workout to workout, the "breathing spells" between exercises until you can do the seven conditioning exercises without resting. You will proceed in the same fashion with the more difficult exercises and increased repetitions at succeeding levels.

You will find the program designed—the progression carefully planned—to make this feasible. You will be able to proceed at your own pace, competing with yourself rather than with anyone else—and this is of great importance for sound conditioning.

Note: Gradually speeding up, from workout to workout, the rate at which you do each exercise will provide greater stimulation for the circulatory and respiratory systems and also help to keep your workouts short. However, the seven conditioning exercises should not be a race against time. Perform each exercise completely to insure maximum benefit.

When and How Often to Work Out

To be most beneficial, exercise should become part of your regular daily routine—as much as bathing, shaving, dressing.

Five workouts a week are called for throughout the program.

You can choose any time that is convenient. Preferably, it should be the same time every day—but it does not matter whether it's first thing in the morning, before dinner in the evening, just before retiring, or any other time.

The hour just before the evening meal is a popular time for exercise. The later afternoon workout provides a welcome change of pace at the end of the work day and helps dissolve the day's worries and tensions.

Another popular time to work out is early morning, before the work day begins. Advocates of the early start say it makes them more alert and energetic on the job.

Among the factors you should consider in developing your workout schedule are personal preference, job and family responsibilities, availability of exercise facilities, and weather. It's important to schedule your workouts for a time when there is little chance that you will have to cancel or interrupt them because of other demands on your time.

You should not exercise strenuously during extremely hot, humid weather or within two hours after eating. Heat and/or digestion both make heavy demands on the circulatory system, and in combination with exercise can be an overtaxing double load.

Your Progress Records

Charts are provided for the orientation program and for each of the five levels.

They list the exercises to be done and the goal for each exercise in terms of number of repetitions, distance, etc.

They also provide space in which to record your progress—(1) in completing the recommended 15 workouts at each level, (2) in accomplishing the three prove-out workouts before moving on to a succeeding level, and (3) in the results as you take the step test from time to time.

A sample chart and progress record for one of the five levels is shown below.

You do the warmup exercises and the conditioning exercises along with one circulatory activity for each workout.

Check off each workout as you complete it. The last three numbers are for the prove-out workouts, in which the seven conditioning exercises should be done without resting. Check them off as you accomplish them.

You are now ready to proceed to the next level.

As you take the step test—at about 2-week intervals—enter your pulse rate.

When you move on to the next level, transfer the last pulse rate from the preceding level. Enter it in the margin to the left of the new progress record and circle it so it will be convenient for continuing reference.

SAMPLE	**GOAL**
Warmup Exercises	**Exercises 1–6 of Orientation Program**
Conditioning Exercises	**Uninterrupted repetitions**
1. Toe touch	20
2. Sprinter	16
3. Sitting stretch	18
4. Pushup	10
5. Situp (fingers laced)	15
6. Leg raiser	16 each leg
7. Flutter kick	40
Circulatory activity (choose one each workout)	
Jog-walk (jog 100, walk 100)	1 mile
Rope (skip 60 secs.; rest 60 secs.)	3 series
Run in place (run 95, hop 15–2 cycles)	3 minutes
Water activities (see page 169)	
Your progress record 1 2 3 4 5 6 7 8 9 10 11 12	**13 14 15**
Step test (pulse)	Prove-out workouts

Getting Set—Orientation Workouts

With the series of preliminary exercises listed in the chart which follows and described on the next two pages, you can get yourself ready—without severe aches or pains—for the progressive conditioning program.

Even if these preliminary exercises should seem easy—and they are deliberately meant to be mild—plan to spend a minimum of one week with them. Do not hesitate to spend two weeks or even three if necessary for you to limber up enough so you can accomplish all the exercises easily and without undue fatigue.

ORIENTATION PROGRAM: MEN	GOAL
Conditioning Exercises	**Repetitions**
*1. Bend and stretch	10
*2. Knee lift	10 left, 10 right
*3. Wing stretcher	20
*4. Half knee bend	10
*5. Arm circles	15 each way
*6. Body bender	10 left, 10 right
7. Prone arch	10
8. Knee pushup	6
9. Head and shoulder curl	5
10. Ankle stretch	15
Circulatory activity (choose one each workout)	
Walking	½ mile
Rope (skip 15 secs.; rest 60 secs.)	3 series

*The first six exercises of the Orientation program will be used as warmup exercises throughout the graded levels.

Step Test Record—After completing the orientation program, take the 2-minute step test. Record your pulse rate here:_____ This will be the base rate with which you can make comparisons in the future.

1. Bend and Stretch

Starting positon: Stand erect, feet shoulder-width apart.
Action: Count 1. Bend trunk forward and down, flexing knees. Stretch gently in attempt to touch fingers to toes or floor. Count 2. Return to starting position.
Note: Do slowly, stretch and relax at intervals rather than in rhythm.

2. Knee Lift

Starting position: Stand erect, feet together, arms at sides.
Action: Count 1. Raise left knee as high as possible, grasping leg with hands and pulling knee against body while keeping back straight. Count 2. Lower to starting position. Counts 3 and 4. Repeat with right knee.

3. Wing Stretcher

Starting position: Stand erect, elbows at shoulder height, fists clenched in front of chest.
Action: Count 1. Thrust elbows backward vigorously without arching back. Keep head erect, elbows at shoulder height. Count 2. Return to starting position.

4. Half Knee Bend

Starting position: Stand erect, hands on hips.
Action: Count 1. Bend knees halfway while extending arms forward, palms down. Count 2. Return to starting position.

5. Arm Circles

Starting positon: Stand erect, arms extended sideward at shoulder height, palms up.
Action: Describe small circles backward with hands. Keep head erect. Do 15 backward circles. Reverse, turn palms down and do 15 small circles forward.

6. Body Bender

Starting position: Stand, feet shoulder-width apart, hands behind neck, fingers interlaced.
Action: Count 1. Bend trunk sideward to left as far as possible, keeping hands behind neck. Count 2. Return to starting position. Counts 3 and 4. Repeat to the right.

7. Prone Arch

Starting position: Lie face down, hands tucked under thighs.
Action: Count 1. Raise head, shoulders, and legs from floor. Count 2. Return to starting position.

8. Knee Pushup

Starting position: Lie on floor, face down, legs together, knees bent with feet raised off floor, hands on floor under shoulders, plams down.
Action: Count 1. Push upper body off floor until arms are fully extended and body is in straight line from head to knees. Count 2. Return to starting position.

9. Head and Shoulder Curl

Starting position: Lie on back, hands tucked under small of back, palms down
Action: Count 1. Tighten abdominal muscles, lift head and pull shoulders and elbows up off floor. Hold for four seconds. Count 2. Return to starting position.

10. Ankle Stretch

Starting position: Stand on a stair, large book or block of wood, with weight on balls of feet and heels raised.
Action: Count 1. Lower heels. Count 2. Raise heels.

Circulatory Activities

WALKING—Step off at a lively pace, swing arms and breathe deeply. *ROPE*—Any form of skipping or jumping is acceptable. Gradually increase the tempo as your skill and condition improve.

LEVEL ONE: MEN   GOAL

Warmup Exercises	Exercises 1–6 of Orientation Program
Conditioning Exercises	**Uninterrupted repetitions**
1. Toe touch	10
2. Sprinter	12
3. Sitting stretch	12
4. Pushup	4
5. Situp (arms extended)	5
6. Leg raiser	12 each leg
7. Flutter kick	30

Circulatory activity (choose one each workout)

Walking (120 steps a minute)	1 mile
Rope (skip 30 secs.; rest 30 secs.)	2 series
Run in place (run 60, hop 10–2 cycles)	2 minutes

Water activities (see page 169)

Your progress record	1 2 3 4 5 6 7 8 9 10 11 12	13 14 15
Step test (pulse)		Prove-out workouts

1. Toe Touch

Starting position: Stand at attention.
Action: Count 1. Bend trunk forward and down keeping knees straight, touching fingers to ankles. Count 2. Bounce and touch fingers to top of feet. Count 3. Bounce and touch fingers to toes. Count 4. Return to starting position.

2. Sprinter

Starting position: Squat, hands on floor, fingers pointed forward, left leg fully extended to rear.
Action: Count 1. Reverse position of feet in bouncing movement, bringing left foot to hands and extending right leg backward—all in one motion. Count 2. Reverse feet again, returning to starting position.

3. Sitting Stretch

Starting position: Sit, legs spread apart, hands on knees.
Action: Count 1. Bend forward at waist, extending arms as far forward as possible. Count 2. Return to starting position.

4. Pushup

Starting position: Lie on floor, face down, legs together, hands on floor under shoulders with fingers pointing straight ahead.
Action: Count 1. Push body off floor by extending arms, so that weight rests on hands and toes. Count 2. Lower the body until chest touches floor.
Note: Body should be kept straight, buttocks should not be raised, abdomen should not sag.

5. Situp (Arms Extended)

Starting position: Lie on back, legs straight and together, arms extended beyond head.

Action: Count 1. Bring arms forward over head, roll up to sitting position, sliding hands along legs, grasping ankles. Count 2. Roll back to starting position.

6. Leg Raiser

Starting position: Right side of body on floor, head resting on right arm.
Action: Lift left leg about 24″ off floor, then lower it. Do required number of repetitions. Repeat on other side.

7. Flutter Kick

Starting position: Lie face down, hands tucked under thighs.
Action: Arch the back, bringing chest and head up, then flutter kick continuously, moving the legs 8—10″ apart. Kick from hips with knees slightly bent. Count each kick as one.

Circulatory Activities

WALKING—Maintain a pace of 120 steps per minute for a distance of 1 mile. Swing arms and breathe deeply.
ROPE—Skip or jump rope continuously using any form for 30 seconds and then rest 30 seconds. Repeat 2 times.
RUN IN PLACE—Raise each foot at least 4″ off floor and jog in place. Count 1 each time left foot touches floor. Complete the number of running steps called for in chart, then do specified number of straddle hops. Complete 2 cycles of alternate running and hopping for time specified on chart.
STRADDLE HOP—*Starting position:* At attention.
Action: Count 1. Swing arms sideward and upward, touching hands above head (arms straight) while simultaneously moving feet sideward and apart in a single jumping motion. Count 2. Spring back to starting position. Two counts in one hop.

LEVEL TWO: MEN	GOAL
Warmup Exercises	**Exercises 1–6 of Orientation Program**
Conditioning Exercises	**Uninterrupted repetitions**
1. Toe touch	20
2. Sprinter	16
3. Sitting stretch	18
4. Pushup	10
5. Situp (fingers laced)	20
6. Leg raiser	16 each leg
7. Flutter kick	40
Circulatory activity (choose one each workout)	
Jog-walk (jog 100; walk 100)	1 mile
Rope (skip 1 min.; rest 1 min.)	3 series
Run in place (run 95, hop 15–2 cycles)	3 minutes
Water activities (see page 169)	

Your progress record	1 2 3 4 5 6 7 8 9 10 11 12	13 14 15
Step test (pulse)		Prove-out workouts

1. Toe Touch

Starting position: Stand at attention.
Action: Count 1. Bend trunk forward and down keeping knees straight, touching fingers to ankles. Count 2. Bounce and touch fingers to top of feet. Count 3. Bounce and touch fingers to toes. Count 4. Return to starting position.

2. Sprinter

Starting position: Squat, hands on floor, fingers pointed forward, left leg fully extended to rear.
Action: Count 1. Reverse position of feet in bouncing movement, bringing left foot to hands, extending right leg backward—all in one motion. Count 2. Reverse feet again, returning to starting position.

3. Sitting Stretch

Starting position: Sit, legs apart, hands on knees.
Action: Count 1. Bend forward at waist, extending arms as far forward as possible. Count 2. Return to starting position.

4. Pushup

Starting position: Lie on floor, face down, legs together, hands on floor under shoulders with fingers pointing straight ahead.
Action: Count 1. Push body off floor by extending arms, so that weight rests on hands and toes. Count 2. Lower the body until chest touches floor.
Note: Body should be kept straight, buttocks should not be raised, abdomen should not sag.

5. Situp (Fingers Laced)

Starting position: Lie on back, legs straight and feet spread approximately 1' apart. Fingers laced behind neck.
Action: Count 1. Curl up to sitting position and turn trunk to left. Touch the right elbow to left knee. Count 2. Return to starting position. Count 3. Curl up to sitting position and turn trunk to right. Touch left elbow to right knee. Count 4. Return to starting position. Score one situp each time you return to starting position. Knees may be bent as necessary.

6. Leg Raiser

Starting position: Right side of body on floor, head resting on right arm.
Action: Lift left leg about 24″ off floor, then lower it. Do required number of repetitions. Repeat on other side.

7. Flutter Kick

Starting position: Lie face down hands tucked under thighs.
Action: Arch the back, bringing chest and head up, then flutter kick continuously, moving the legs 8—10″ apart. Kick from hips with knees slightly bent. Count each kick as one.

Circulatory Activities

JOG-WALK—Jog and walk alternately for number of paces indicated on chart for distance specified.
ROPE—Skip or jump rope continuously using any form for 60 seconds and then rest 60 seconds. Repeat 5 times.
RUN IN PLACE—Raise each foot at least 4″ off floor and jog in place. Count 1 each time left foot touches floor. Complete number of running steps called for in chart, then do specified number of straddle hops. Complete 2 cycles of alternate running and hopping for time specified on chart.
STRADDLE HOP—Starting position: at attention.
Action: Count 1. Swing arms sideward and upward, touching hands above head (arms straight) while simultaneously moving feet sideward and apart in a single jumping motion. Count 2. Spring back to starting position. Two counts in one hop.

LEVEL THREE: MEN	GOAL
Warmup Exercises	**Exercises 1–6 of Orientation Program**
Conditioning Exercises	**Uninterrupted repetitions**
1. Toe touch	30
2. Sprinter	20
3. Sitting stretch (fingers laced)	18
4. Pushup	20
5. Situp (arms extended, knees up)	30
6. Leg raiser	20 each leg
7. Flutter kick	50
Circulatory activity (choose one each workout)	
Jog-walk (jog 200; walk 100)	1½ miles
Rope (skip 1 min.; rest 1 min.)	5 series
Run in place (run 135, hop 20–2 cycles)	4 minutes
Water activities (see page 169)	

Your progress record	1 2 3 4 5 6 7 8 9 10 11 12	13 14 15
Step test (pulse)		Prove-out workouts

1. Toe Touch

Starting position: Stand at attention.
Action: Count 1. Bend trunk forward and down keeping knees straight, touching fingers to ankles. Count 2. Bounce and touch fingers to top of feet. Count 3. Bounce and touch fingers to toes. Count 4. Return to starting position.

2. Sprinter

Starting position: Squat, hands on floor, fingers pointed forward, left leg fully extended to rear.
Action: Count 1. Reverse position of feet in bouncing movement, bringing left foot to hands, extending right leg backward—all in one motion. Count 2. Reverse feet again, returning to starting position.

3. Sitting Stretch (Fingers Laced)

Starting position: Sit, legs spread apart, fingers laced behind neck, elbows back.
Action: Count 1. Bend forward at waist, reaching elbows as close to floor as possible. Count 2. Return to starting position.

4. Pushup

Starting position: Lie on floor, face down, legs together, hands on floor under shoulders with fingers pointing straight ahead.
Action: Count 1. Push body off floor by extending arms, so that weight rests on hands and toes. Count 2. Lower the body until chest touches floor.
Note: Body should be kept straight, buttocks should not be raised, abdomen should not sag.

5. Situp (Arms Extended, Knees Up)

Starting position: Lie on back, legs straight, arms extended overhead.
Action: Count 1. Sit up, reaching forward with arms encircling knees while pulling them tightly to chest. Count 2. Return to starting position. Do this exercise rhythmically, without breaks in the movement.

6. Leg Raiser

Starting position: Right side of body on floor, head resting on right arm.
Action: Lift left leg about 24" off floor then lower it. Do required number of repetitions. Repeat on other side.

7. Flutter Kick

Starting position: Lie face down, hands tucked under thighs.
Action: Arch the back, bringing chest and head up, then flutter kick continuously, moving the legs 8–10" apart. Kick from hips with knees slightly bent. Count each kick as one.

Circulatory Activities

JOG-WALK—Jog and walk alternately for number of paces indicated on chart for distance specified.
ROPE—Skip or jump rope continuously using any form for 60 seconds and then rest 60 seconds. Repeat 5 times.
RUN IN PLACE—Raise each foot at least 4" off floor and jog in place. Count 1 each time left foot touches floor. Complete number of running steps called for in chart, then do specified number of straddle hops. Complete 2 cycles of alternate running and hopping for time specified on chart.
STRADDLE HOP—Starting position: At attention.
Action: Count 1. Swing arms sideward and upward, touching hands above head (arms straight) while simultaneously moving feet sideward and apart in a single jumping motion. Count 2. Spring back to starting position. Two counts in one hop.

LEVEL FOUR: MEN   **GOAL**

Warmup Exercises	Exercises 1–6 of Orientation Program
Conditioning Exercises	**Uninterrupted repetitions**
1. Toe touch (twist and bend)..	20 each side
2. Sprinter	28
3. Sitting stretch (alternate)	24
4. Pushup..	30
5. Situp (arms crossed, knees bent).................................	30
6. Leg raiser (whip)	20 each leg
7. Prone arch (arms extended)	20
Circulatory activity (choose one each workout)	
Jog	1 mile
Rope (skip 90 secs.; rest 30 secs.)	3 series
Run in place (run 180; hop 25–2 cycles)	5 minutes
Water activities (see page 169)	

Your progress record	1 2 3 4 5 6 7 8 9 10 11 12	13 14 15
Step test (pulse)		Prove-out workouts

1. Toe Touch
(Twist and Bend)

Starting position: Stand, feet shoulder-width apart, arms extended over head, thumbs interlocked.
Action: Count 1. Twist trunk to right and touch floor inside right foot with fingers of both hands. Count 2. Touch floor outside toes of right foot. Count 3. Touch floor outside heel of right foot. Count 4. Return to starting position, sweeping trunk and arms upward in a wide arc. On the next four counts, repeat action to left side.

2. Sprinter

Starting position: Squat, hands on floor, fingers pointed forward, left leg fully extended to rear.
Action: Count 1. Reverse position of feet in bouncing movement, bringing left foot to hands, extending right leg backward—all in one motion. Count 2. Reverse feet again, returning to starting position.

3. Sitting Stretch (Alternate)

Starting position: Sit, legs spread apart, fingers laced behind neck, elbows back.

Action: Count 1. Bend foreward to left, touching forehead to left knee. Count 2. Return to starting position. Counts 3 and 4. Repeat to right. Score one repetition each time you return to starting position. Knees may be bent if necessary.

4. Pushup

Starting position: Lie on floor, face down, legs together, hands on floor under shoulders with fingers pointing straight ahead.
Action: Count 1. Push body off floor by extending arms, so that weight rests on hands and toes. Count 2. Lower the body until chest touches floor.
Note: Body should be kept straight, buttocks should not be raised, abdomen tight.

5. Situp
(Arms Crossed, Knees Bent)

Starting position: Lie on back, arms crossed on chest, hands grasping opposite shoulders, knees bent to right angle, feet flat on floor.
Action: Count 1. Curl up to sitting position. Count 2. Return to starting position.

6. Leg Raiser (Whip)

Starting position: Right side of body on floor, right arm supporting head.
Action: Whip left leg up and down rapidly, lifting as high as possible off the floor. Count each whip as one. Reverse position and whip right leg up and down.

7. Prone Arch (Arms Extended)

Starting position: Lie face down, legs straight and together, arms extended to sides at shoulder level.
Action: Count 1. Arch the back, bringing arms, chest and head up, and raising legs as high as possible. Count 2. Return to starting position.

Circulatory Activities

JOG—Jog continuously for 1 mile.
ROPE—Skip or jump rope continuously using any form for 90 seconds and then rest for 30 seconds. Repeat 3 times.
RUN IN PLACE—Raise each foot at least 4" off floor and jog in place. Count 1 each time left foot touches floor. Complete number of running steps called for in chart, then do specified number of straddle hops. Complete 2 cycles of alternate running and hopping in time specified on chart.
STRADDLE HOP—Starting position: At attention.
Action: Count 1. Swing arms sideward and upward, touching hands above head (arms straight) while simultaneously moving feet sideward and apart in a single jumping motion. Count 2. Spring back to starting position. Two counts in one hop.

LEVEL FIVE: MEN · GOAL

Warmup Exercises	Exercises 1–6 of Orientation Program
Conditioning Exercises	**Uninterrupted repetitions**
1. Toe touch (twist and bend)	30 each side
2. Sprinter	36
3. Sitting stretch (alternate)	30
4. Pushup	50
5. Situp (fingers laced, knees bent)	40
6. Leg raiser (on extended arm)	20 each side
7. Prone arch (fingers laced)	30
Circulatory activity (choose one each workout)	
Jog-run	3 miles
Rope (skip 2 mins.; rest 30 secs.)	3 series
Run in place (run 216; hop 30–2 cycles)	6 minutes
Water activities (see page 169)	

Your progress record	1 2 3 4 5 6 7 8 9 10 11 12	13 14 15
Step test (pulse)		Prove-out workouts

1. Toe Touch (Twist and Bend)

Starting position: Stand, feet shoulder-width apart, arms extended over head, thumbs interlocked.
Action: Count 1. Twist trunk to right and touch floor inside right foot with fingers of both hands. Count 2. Touch floor outside toes of right foot. Count 3. Touch floor outside heel of right foot. Count 4. Return to starting position, sweeping trunk and arms upward in a wide arc. On the next four counts, repeat action to left side.

2. Sprinter

Starting position: Squat, hands on floor, fingers pointed forward, left leg fully extended to rear.

Action: Count 1. Reverse position of feet in bouncing movement, bringing left foot to hands and extending right leg backward—all in one motion. Count 2. Reverse feet again, returning to starting position.

3. Sitting Stretch (Alternate)

Starting position: Sit, legs spread apart, fingers laced behind neck, elbows back.

Action: Count 1. Bend forward to left, touching forehead to left knee. Count 2. Return to starting position. Counts 3 and 4. Repeat to right. Score one repetition each time you return to starting position. Knees may be bent if necessary.

4. Pushup

Starting position: Lie on floor, face down, legs together, hands on floor under shoulders with fingers pointing straight ahead.

Action: Count 1. Push body off floor by extending arms so that weight rests on hands and toes. Count 2. Lower body until chest touches floor.

Note: Body should be kept straight, buttocks should not be raised, abdomen should not sag.

5. Situp
(Fingers Laced, Knees Bent)

Starting position: Lie on back, fingers laced behind neck, knees bent, feet flat on floor.

Action: Count 1. Sit up, turn trunk to right, touch left elbow to right knee. Count 2. Return to starting position. Count 3. Sit up, turn trunk to left, touch right elbow to left knee. Count 4. Return to starting position. Score one each time you return to starting position.

6. Leg Raiser
(On Extended Arm)

Starting position: Body rigidly supported by extended right arm and foot. Left arm is held behind head.

Action: Count 1. Raise left leg high. Count 2. Return to starting position slowly. Do required number of repetitions. Repeat on other side.

7. Prone Arch
(Fingers Laced)

Starting position: Lie face down, fingers laced behind neck.

Action: Count 1. Arch back, legs, and chest off floor. Count 2. Extend arms forward. Count 3. Return hands to behind neck. Count 4. Flatten body to floor.

Circulatory Activities

JOG RUN—Alternately jog and run the specified distance. Attempt to increase the proportion of time spent running in each successful workout.

ROPE—Skip or jump rope continuously using any form for 2 minutes and then rest 30 seconds. Repeat 3 times.

RUN IN PLACE—Raise each foot at least 4″ off floor and jog in place. Count 1 each time left foot touches floor. Complete number of running steps called for in chart, then do specified number of straddle hops. Complete 2 cycles of alternate running and hopping for time specified on the chart.

STRADDLE HOP—*Starting position:* At attention.

Action: Count 1. Swing arms sideward and upward, touching hands above head (arms straight) while simultaneously moving feet sideward and apart in a single jumping motion. Count 2. Spring back to starting position. Two counts in one hop.

(End of program for men, see pages 167–170 for more instruction.)

Staying Fit

Once you have reached the level of conditioning you have chosen for yourself, you will wish to maintain your fitness.

To do so, continue the workouts at that level.

While it is possible to maintain fitness with three workouts a week, ideally, exercise should be a daily habit. If you can, by all means continue your workouts on a five-times-a-week basis.

If at any point—either after reaching your goal or in the process of doing so—your workouts are interrupted because of illness or other reasons for more than a week, it will be best to begin again at a lower level. If you have had a serious illness or surgery, proceed under your physician's guidance.

Broadening Your Program

The exercises and activities you have engaged in are basic—designed to take you soundly and progressively up the ladder to physical fitness without need for special equipment or facilities.

There are many other activities and forms of exercise which, if you wish, you may use to supplement the basic program.

They include a variety of sports; water exercises you can use if you have access to a pool; and isometrics—sometimes called exercises without movement—which take little time (6–8 seconds each).

Isometrics

Isometric contraction exercises take very little time, and require no special equipment. They're excellent muscle strengtheners and, as such, valuable supplements.

The idea of isometrics is to work out a muscle by pushing or pulling against an immovable object such as a wall . . . or by pitting it against the opposition of another muscle.

The basis is the "overload" principle of exercise physiology—which holds that a muscle required to perform work beyond the usual intensity will grow in strength. And research has indicated that one hard, 6 to 8-second isometric contraction per workout can, over a period of six months, produce a significant strength increase in a muscle.

The exercises illustrated and described on the following page cover major large muscle groups of the body.

They can be performed almost anywhere and at almost any time.

There is no set order for doing them—nor do all have to be completed at one time. You can, if you like, do one or two in the morning, others at various times during the day whenever you have half a minute or even less to spare.

For each contraction, maintain tension *no more than eight seconds*. Do little breathing during a contraction; breathe deeply between contractions.

And start easily. Do *not* apply maximum effort in the beginning.

For the first three or four weeks, you should exert only about one-half what you think is your maximum force.

Use the first three or four seconds to build up to this degree of force—and the remaining four or five seconds to hold it.

For the next two weeks, gradually increase force to more nearly approach maximum. After about six weeks, it will be safe to exert maximum effort.

Pain indicates you're applying too much force; reduce the amount immediately. If pain continues to accompany any exercise, discontinue using that exercise for a week or two. Then try it again with about 50 percent of maximum effort and, if no pain occurs, you can go on to gradually build up toward maximum.

Neck

Starting position: Sit or stand, with interlaced fingers of hands on forehead.
Action: Forcibly exert a forward push of head while resisting equally hard with hands.

Starting position: Sit or stand, with interlaced fingers of hands behind head.
Action: Push head backward while exerting a forward pull with hands.

Starting position: Sit or stand, with palm of left hand on left side of head.
Action: Push with left hand while resisting with head and neck. Reverse using right hand on right side of head.

Upper Body

Starting position: Stand, back to wall, hands at sides, palms toward wall.
Action: Press hands backward against wall, keeping arms straight.

Starting position: Stand, facing wall, hands at sides, palms toward wall.
Action: Press hands forward against wall, keeping arms straight.

Starting position: Stand in doorway or with side against wall, arms at sides, palms toward legs.
Action: Press hand(s) outward against wall or doorframe, keeping arms straight.

Arms

Starting position: Stand with feet slightly apart. Flex right elbow, close to body, palm up. Place left hand over right.
Action: Forcibly attempt to curl right arm upward, while giving equally strong resistance with the left hand. Repeat with left arm.

Arms and Chest

Starting position: Stand with feet comfortably spaced, knees slightly bent. Clasp hands, palms together, close to chest.
Action: Press hands together and hold.

Starting position: Stand with feet slightly apart, knees slightly bent. Grip fingers, arms close to chest.
Action: Pull hard and hold.

Abdominal

Starting position: Stand, knees slightly flexed, hands resting on knees.
Action: Contract abdominal muscles.

Lower Back, Buttocks and Back of Thighs

Starting position: Lie face down, arms at sides, palms up, legs placed under bed or other heavy object.
Action: With both hips flat on floor, raise one leg, keeping knee straight so that heel pushes hard against the resistance above. Repeat with opposite leg.

Legs

Starting position: Sit in chair with left ankle crossed over right, feet resting on floor, legs bent at 90 degree angle.
Action: Forcibly attempt to straighten right leg while resisting with the left. Repeat with opposite leg.

Inner and Outer Thighs

Starting position: Sit, legs extended with each ankle pressed against the outside of sturdy chair legs.
Action: Keep legs straight and pull toward one another firmly. For outer thigh muscles, place ankles inside chair legs and exert pressure outward.

Water Activities

Swimming is one of the best physical activities for people of all ages.

With the body submerged in water, blood circulation automatically increases to some extent; pressure of water on the body also helps promote deeper ventilation of the lungs; and with well-planned activity, both circulation and ventilation increase still more.

The water exercises described below can be used either as supplements to, or replacements for, the circulatory activities of the basic program. The goals for each of the five levels are shown in the chart below.

Level	1	2	3	4	5
Bobs	10	15	25	75	125
Swim	5 min	10 min	15 min	—	—
Interval swimming	—	—	—	25 yds. (Repeat 20 times.)	50 yds. (Repeat 20 times.)

Bobbing

Starting position: Face out of water.

Action: Count 1. Take a breath. Count 2. Submerge while exhaling until feet touch bottom. Count 3. Push up from bottom to surface while continuing to exhale. Three counts to one bob.

Swimming

Use any type of stroke. Swim continuously for the time specified.

Interval Swimming

Use any type of stroke. Swim moderately fast for distance specified. You can then either swim back slowly to starting point or get out of pool and walk back. Repeat specified number of times.

Weight Training

Weight training also is an excellent method of developing muscular strength—and muscular endurance. Where equipment is available, it may be used as a supplement to the seven conditioning exercises.

Because of the great variety of weight training exercises, there will be no attempt to describe them here. Both barbells and weighted dumbbells—complete with instructions—are available at most sporting goods stores. A good rule to follow in deciding the maximum weight you should lift is to select a weight you can lift six times without strain.

Sports

Soccer, basketball, handball, squash, ice hockey, and other sports that require sustained effort can be valuable aids to building circulatory endurance.

But if you have been sedentary, it's important to pace yourself carefully in such sports, and it may even be advisable to avoid them until you are well along in your physical conditioning program. That doesn't mean you should avoid all sports.

There are many excellent conditioning and circulatory activities in which the amount of exertion is easily controlled and in which you can progress at your own rate. Bicycling is one example. Others include hiking, skating, tennis, running, cross-country skiing, rowing, canoeing, water skiing.

You can engage in these sports at any point in the program, if you start slowly. Games should be played with full speed and vigor only when your conditioning permits doing so without undue fatigue.

On days when you get a good workout in sports, you can skip part or all of your exercise program. Use your own judgment.

If you have engaged in a sport which exercises the legs and stimulates the heart and lungs—such as skating—you could skip the circulatory activity for that day, but you still should do some of the conditioning and stretching exercises for the upper body. On the other hand, weight-lifting is an excellent conditioning activity, but it should be supplemented with running or one of the other circulatory exercises.

Whatever your favorite sport, you will find your enjoyment enhanced by improved fitness.

PART FOUR

OTHER INFORMATION

DEALING WITH EMERGENCIES

A sanitation worker is not a police officer, a firefighter, a doctor, or an emergency services technician, nor is the sanitation worker expected to perform the duties of any of these. However, the sanitation worker is often in a position to spot emergencies before others know of them. The seat in the cab of most sanitation vehicles is high above the street. This high perch gives the sanitation worker a good view of what is happening all around the area. Also, the sanitation vehicle is often on the streets early in the morning before many people are about and operating in weather conditions when few other people are out. The sanitation worker might be the first to see smoke coming from a window, an automobile accident, or a victim of illness or violence lying in the gutter.

Work on heavy sanitation equipment is also not without its dangers. Sanitation workers can fall from moving vehicles, get caught in machinery, fall over objects in their paths or be hurt by materials that they are moving.

No sanitation worker is expected to single-handedly deal with such emergencies, but each does have some responsibility. The first responsibility of the sanitation worker who notices an emergency situation is to summon help. If the sanitation vehicle is equipped with two-way communications, the worker must use these. If not, the worker must seek out the nearest fire or police call box or public telephone. If no public communication facility is in sight, the sanitation worker must enlist the help of the nearest private citizen. The point is to immediately notify police, fire services, or emergency medical services of the location and the nature of the emergency.

The responsibility of the sanitation worker does not include putting his or her own life in danger. Common sense says to stay out of the line of fire of criminal activity in progress and not to enter a building heavily engulfed in flames. On the other hand, the sanitation worker should, after sending for professional assistance, give first aid and emergency assistance to the extent to which he or she is capable.

Some of the most common emergencies with which a sanitation worker is likely to come in contact are: electrical shock; fire; breathing stoppage; heavy bleeding; broken bones; and the consequences of falls.

Obviously, if a sanitation worker sees a dangling wire that might possibly be a live electric wire, the worker should not touch that wire. He or she should notify police or supervisors so that the electric company can be alerted. Sometimes, though, a person or piece of equipment may come into contact with a live electric wire which was unnoticed. In case of electric shock, the first thing to do is to remove the person from contact with the electricity. If the source of the power is known, shut it off. If there is no handy switch the victim must be separated from the electricity. This must be done with great caution. The human body conducts electricity. This means that the rescuer who touches a victim still in contact with live electricity will have the electricity transmitted to him or herself. The shock victim must be pushed away from the electric source with a nonmetallic instrument such as a board, a rubber tire or heavy rope. Never touch an electric shock victim until he or she has been fully separated from the electric source.

A victim of electric shock may have burns and may have stopped breathing. Burns should be covered loosely with clean cloths; nothing more. Breathing stoppage requires artificial respiration.

Mouth-to-mouth resuscitation is today's most commonly used method of artificial respiration. It is simple enough to do without training or practice. Here are the steps:

1. Lay the victim on his or her back. Tilt the head back so that the chin points up. Clear the airway of any obstructions. Check to be certain that the person really is not breathing by watching for movement of the chest and by putting your ear against the mouth to listen for or feel exhaled breath.
2. If there are no signs of the victim's breathing.
   a) keep the victim's head tilted back and pinch the nose shut,
   b) take a deep breath,
   c) cover the victim's mouth tightly with your own open mouth and exhale into the victim's mouth. After exhaling two full breaths into the victim's mouth, remove your own mouth so that the victim can exhale.
   d) Watch for signs that victim is breathing independently. If the chest does not rise after the victim exhales, repeat the mouth-to-mouth breathing every 5 seconds (about 12 times a minute) until help arrives or until the victim begins breathing.
   e) If the victim is a child, cover both nose and mouth with your mouth and make more frequent, shallower breaths.

It is a good idea to take a formal course in mouth-to-mouth resuscitation.

Some burns are caused by electricity and some by chemicals; however, the most common cause of burns is fire. If you see a major fire in progress—in a house, an automobile, or a large open field or lot—summon firefighters at once. If you see a small fire—the beginning of an engine fire in your vehicle, a small fire in refuse at roadside, or the beginnings of a fire in your cargo area—put it out. You should have a fire extinguisher or two on your vehicle. Learn what type of extinguisher(s) you carry and the instructions for use before you ever need to use an extinguisher. Water can be used to put out wood, paper, or cloth fires or to cool a burning tire, but pouring water on most other fires can be dangerous. Water on a gasoline fire will just spread the flames. Water on an electrical fire creates a shock hazard. If you have a fire extinguisher, it is safer to use it. Use a B:C type of extinguisher on electrical fires and burning liquids. Use the A:B:C type for burning wood, paper, and cloth. Know what kind of extinguisher you have and know how it works. Study the instructions printed on the extinguisher before you set out in a new vehicle.

- Aim at the base of the fire, not into the flames.
- Stay as far from the fire as possible while using the extinguisher.
- Stand upwind of the fire. Let the wind blow the contents of the extinguisher into the fire, not the fire to the extinguisher.
- Use up the contents of the extinguisher even if you think that the fire is out. The fire could flare up again, and the extinguisher cannot be reused without recharging anyway.

A fire extinguisher may not be the best solution if a person is on fire. The best way to put out a fire on a person—whether it be the person's clothing or hair—is by smothering the fire. Wrap the person in blankets, coats or a rug and roll the person over and over until the fire is out. Then keep the person lying down and loosely covered and send for medical help. Do not try to treat the burns yourself; do not try to make the person more comfortable by offering food or drink. Watch to make sure the person continues breathing and give words of comfort and reassurance if the person is conscious, but wait for a trained professional to give medical help.

Severe bleeding may require more active assistance. An injured person may rapidly bleed to death if a major artery is severed. Therefore, along with sending for help, you must try to slow the bleeding. A person who is losing a lot of blood is likely to faint. So, have the person lie down. Raise the area of bleeding above the height of the heart if at all possible, cover the wound with a cloth (clean, if available), and apply firm, direct pressure. This may be unpleasant for you and painful for the victim, but you may be saving a life.

The other emergency a sanitation worker is likely to meet is broken bones, most often from falls. You can use a little common sense when dealing with broken bones. A person with a broken arm or hand can walk out of harm's way and wait for transportation to medical services by car or other vehicle. A person who may have broken bones in a foot by dropping a heavy object can hop to the side of the street to await help. On the other hand, the victim who has fallen from the truck and is lying in the street should not be moved. Whenever the head, neck, or back may have been injured, there is the real danger of spinal cord injury. A severe spinal cord injury can leave the victim crippled for life. Only a trained person should move the victim if there is any possibility at all of spinal cord injury. If a coworker is hurt in this way, the best assistance you can give is to serve as a director of traffic. Leave the victim lying in the street, covered lightly, and direct traffic around the accident area. Send for paramedics who have equipment to move the victim without causing further injury.

Your sanitation worker exam may ask a few questions about dealing with emergencies or about first aid knowledge. Use your common sense and this information in choosing the best answer to these questions. The sanitation worker should remember not to step in where he or she is unqualified and may do more harm than good. However, if the greater harm would come from doing nothing, then the sanitation worker should do whatever he or she can to help.

VEHICLE SAFETY

The sanitation worker is not expected to be a trained mechanic; however, he or she is expected to be aware of the condition of the vehicle and to take responsibility for reporting problems or dangerous conditions.

In some localities, the same driver operates the same vehicle every day. This driver gets to know the vehicle well and can quickly recognize when it doesn't handle, sound, feel, or smell just right. The driver who takes out the same vehicle each day may be personally responsible for minor adjustments and even for routine care. Daily reporting may not be required. This driver may just bring the vehicle for regularly scheduled maintenance and when there is a problem.

In other localities, drivers are assigned to vehicles on a daily basis. A careful morning checkout of the vehicle is more important to the driver who is using an unfamiliar vehicle. If a driver does not know the special "feel" of a particular vehicle, he or she must be even more careful to check for potential problems. Likewise, the driver who turns in a vehicle in the evening that someone else may drive the next day has the extra responsibility of filing a complete and accurate report on its condition. That driver must make a turn-in inspection and fill out whatever reports are required.

Check-out and check-in inspections make sense in terms of personal safety of drivers and crews and in terms of maintaining the life of sanitation vehicles. Even where not required by department rules these are a good idea.

Whether or not you are required to do daily safety inspections by your department, you will have to learn to do these to pass the Class B Commercial Driver's License Test (CDL). There are far more drivers of tractor trailers, tank trucks, double trailers, and other heavy-duty, long-distance haulers than there are drivers of sanitation vehicles. The CDL was drafted to apply to all of them. Daily inspections are much more necessary and much more meaningful when the vehicle may have covered 600 miles that day. But, since you have to take the same test that those drivers have to take, you may as well learn what you can about daily vehicle inspection.

The first part of your inspection is done without touching the vehicle at all. Basically, you look. Look under the vehicle. Is there a puddle? Use your eyes and your nose. A gasoline leak will be close to the gasoline tank and will smell distinctly like gasoline. Antifreeze/coolant is green. Motor oil is brown and is thick and greasy. Transmision fluid is also oily but is pinkish and is a thinner oil.

After looking for puddles, look at wheels and tires. Be sure all tires are properly inflated. Mechanics should be monitoring tread depth for you, but you can double check. Also look at the tire sidewalls. Steel cords or fabric popping out at the sidewalls is very dangerous. You should not drive at all on tires with worn or damaged sidewalls. Be sure all lugs are in place and tight and that rims are not bent or damaged. And take a look at valve stems to be sure that they are in good condition. If even one tire leaks air while you are on the road, you will find maneuvering more difficult.

Don't crawl under the vehicle, but take a look at the tail pipe and other visible parts of the exhaust system to be sure that the pipe is not missing, broken or compressed.

As you walk around the vehicle checking for leaks and tire condition, notice the lights. If the glass is dirty, wash the headlamps, taillights, signals, flashers, reflectors, etc. You need to have clean lights in order to see and to be seen. This is such a simple safety measure, yet it really pays off.

As long as you are wiping or washing off the lights, hop up and do the same for mirrors and windshield. As you can see, much of the daily inspection is really based on common sense.

Obviously, greater visibility leads to greater safety. While looking at the windshield, check the wiper blades to be certain that they are secure and that they make contact with the windows throughout their length.

As you walk around the vehicle, you may notice body damage or hairline cracks or tiny holes in windshield or other glass. Mark these down. You don't want it on your record that you damaged a vehicle that you found already damaged. And tiny breaks or holes in the windshield can be repaired before they become so large that the whole piece of glass needs to be replaced. Bring small problems to the attention of the supervisor or mechanics so that they can be attended to at the next regularly scheduled overhaul or even sooner if necessary.

Now get into the cab or driver's seat. Try the horn. Turn on lights one-by-one and make certain that all of them work. You may need someone to stand behind the vehicle to tell you whether your brake lights and backup lights are working properly. Turn on the windshield wipers and try the washers. Everything should work. Start the engine and listen for strange noises. Look at all the gages and warning lights. Some of these may not be operative until the engine has been running for awhile or until the vehicle is in motion, but watch all those that might indicate problems in a stationary vehicle. With the engine running, you can test the brakes for pressure even if not for stopping speed or balance. You should also check the steering wheel to be sure it does not have too much play. You want the steering wheel to be responsive when you turn it.

Before you get out of the vehicle, check to be certain that fire extinguishers, safety triangles or flares, extra fuses, and any other required safety devices are in place. Look into the first aid kit and make sure all supplies are in place. If any items in the first aid kit have been used, resupply before starting out. Finally, open and shut all doors. Make sure that they latch securely.

The next place to look is under the hood or in the engine compartment. There are many safety checks you can make here without any formal mechanical training. First, check fluid levels. The fluid levels you can easily check are:

- motor oil
- transmission fluid
- brake fluid
- power steering fluid
- radiator coolant
- windshield washer fluid
- battery level

Next look at belts and hoses. Belts and hoses must be securely attached and must be neither stiff and rigid nor soft and spongy. There must be no nicks or cuts which could lead to breakage on the road. If you are not certain as to just how belts and hoses should feel, ask a supervisor. There is a very distinct feel to dried-out rubber and to rotting rubber. Once you have been shown, you will know how firm but flexible belts and hoses feel.

Also in the engine compartment you should look out for loose electrical connections, frayed wires, or buildup of deposits of any sort. If you notice any wires or parts that seem wobbly or out of line, ask for a second opinion. You are not expected to be an expert; you are expected to be alert to potential problems or safety hazards.

Your responsibility for vehicle safety does not end when you leave the garage. While driving, you must keep your eyes open for irregularities on any of the gages. Listen for unusual sounds. Notice if the steering is not responsive, if shifting is difficult, if acceleration is not as expected, if brakes grab unevenly, or if there is excessive pull. Stop on a hill and check to be sure that the emergency brake holds the vehicle securely. While driving and when you stop, use your nose—gasoline leaks announce themselves by their smell. Burning rubber, also, has a strong, distinctive smell. Burning rubber may come from double tires that are touching, from a belt that is travelling badly, from faulty transmission, or from brake problems. Bring the vehicle back in for service if you smell burning rubber. Carbon monoxide, the most dangerous component of exhaust has no smell; however, gasoline exhaust fumes do have an odor. If you smell exhaust inside your moving vehicle, you must suspect a problem with the exhaust system. If this is the case, you are breathing in carbon monoxide along with the other exhaust fumes. The exhaust system of this vehicle must be checked out and repaired very soon.

Each type of vehicle has special safety requirements related to its use. If your vehicle is a collection truck, for instance, give special attention to hand- and footholds used by the crew. These must be tightly bolted to the vehicle and free of grease that might cause workers to slip and fall off. Hopper-operating machinery must also be in good working order so that no one gets caught and injured in it. If the vehicle is a street sweeper, brushes must easily clear the tires so as not to rub away at them. Snow plow blades must raise freely so that the plow will not be abruptly stopped and overturned by steel plates in the roadway. Salt and sand spreaders must not reverse and spray material into the driver's line of vision.

Sanitation workers also have special concerns during the daily operation of their vehicles. All the sanitation features of the vehicle must remain in top condition at all times. The dangers mentioned in the previous paragraph can appear at any time. Sanitation workers must be constantly alert to machinery malfunctions and to weather-created hazards like ice buildup on steps and handholds, glazed windshields, and frozen windshield wipers.

In short, vehicle safety is a full time concern—for your departmental written exam, for the CDL exam, when you start out each day, and throughout the working day as well.

EXTRA ANSWER SHEETS FOR EXTRA
PRACTICE WITH MODEL EXAMS

1. Ⓐ Ⓑ Ⓒ Ⓓ   16. Ⓐ Ⓑ Ⓒ Ⓓ   31. Ⓐ Ⓑ Ⓒ Ⓓ   46. Ⓐ Ⓑ Ⓒ Ⓓ

2. Ⓐ Ⓑ Ⓒ Ⓓ   17. Ⓐ Ⓑ Ⓒ Ⓓ   32. Ⓐ Ⓑ Ⓒ Ⓓ   47. Ⓐ Ⓑ Ⓒ Ⓓ

3. Ⓐ Ⓑ Ⓒ Ⓓ   18. Ⓐ Ⓑ Ⓒ Ⓓ   33. Ⓐ Ⓑ Ⓒ Ⓓ   48. Ⓐ Ⓑ Ⓒ Ⓓ

4. Ⓐ Ⓑ Ⓒ Ⓓ   19. Ⓐ Ⓑ Ⓒ Ⓓ   34. Ⓐ Ⓑ Ⓒ Ⓓ   49. Ⓐ Ⓑ Ⓒ Ⓓ

5. Ⓐ Ⓑ Ⓒ Ⓓ   20. Ⓐ Ⓑ Ⓒ Ⓓ   35. Ⓐ Ⓑ Ⓒ Ⓓ   50. Ⓐ Ⓑ Ⓒ Ⓓ

6. Ⓐ Ⓑ Ⓒ Ⓓ   21. Ⓐ Ⓑ Ⓒ Ⓓ   36. Ⓐ Ⓑ Ⓒ Ⓓ   51. Ⓐ Ⓑ Ⓒ Ⓓ

7. Ⓐ Ⓑ Ⓒ Ⓓ   22. Ⓐ Ⓑ Ⓒ Ⓓ   37. Ⓐ Ⓑ Ⓒ Ⓓ   52. Ⓐ Ⓑ Ⓒ Ⓓ

8. Ⓐ Ⓑ Ⓒ Ⓓ   23. Ⓐ Ⓑ Ⓒ Ⓓ   38. Ⓐ Ⓑ Ⓒ Ⓓ   53. Ⓐ Ⓑ Ⓒ Ⓓ

9. Ⓐ Ⓑ Ⓒ Ⓓ   24. Ⓐ Ⓑ Ⓒ Ⓓ   39. Ⓐ Ⓑ Ⓒ Ⓓ   54. Ⓐ Ⓑ Ⓒ Ⓓ

10. Ⓐ Ⓑ Ⓒ Ⓓ   25. Ⓐ Ⓑ Ⓒ Ⓓ   40. Ⓐ Ⓑ Ⓒ Ⓓ   55. Ⓐ Ⓑ Ⓒ Ⓓ

11. Ⓐ Ⓑ Ⓒ Ⓓ   26. Ⓐ Ⓑ Ⓒ Ⓓ   41. Ⓐ Ⓑ Ⓒ Ⓓ   56. Ⓐ Ⓑ Ⓒ Ⓓ

12. Ⓐ Ⓑ Ⓒ Ⓓ   27. Ⓐ Ⓑ Ⓒ Ⓓ   42. Ⓐ Ⓑ Ⓒ Ⓓ   57. Ⓐ Ⓑ Ⓒ Ⓓ

13. Ⓐ Ⓑ Ⓒ Ⓓ   28. Ⓐ Ⓑ Ⓒ Ⓓ   43. Ⓐ Ⓑ Ⓒ Ⓓ   58. Ⓐ Ⓑ Ⓒ Ⓓ

14. Ⓐ Ⓑ Ⓒ Ⓓ   29. Ⓐ Ⓑ Ⓒ Ⓓ   44. Ⓐ Ⓑ Ⓒ Ⓓ   59. Ⓐ Ⓑ Ⓒ Ⓓ

15. Ⓐ Ⓑ Ⓒ Ⓓ   30. Ⓐ Ⓑ Ⓒ Ⓓ   45. Ⓐ Ⓑ Ⓒ Ⓓ   60. Ⓐ Ⓑ Ⓒ Ⓓ

1. Ⓐ Ⓑ Ⓒ Ⓓ	16. Ⓐ Ⓑ Ⓒ Ⓓ	31. Ⓐ Ⓑ Ⓒ Ⓓ	46. Ⓐ Ⓑ Ⓒ Ⓓ
2. Ⓐ Ⓑ Ⓒ Ⓓ	17. Ⓐ Ⓑ Ⓒ Ⓓ	32. Ⓐ Ⓑ Ⓒ Ⓓ	47. Ⓐ Ⓑ Ⓒ Ⓓ
3. Ⓐ Ⓑ Ⓒ Ⓓ	18. Ⓐ Ⓑ Ⓒ Ⓓ	33. Ⓐ Ⓑ Ⓒ Ⓓ	48. Ⓐ Ⓑ Ⓒ Ⓓ
4. Ⓐ Ⓑ Ⓒ Ⓓ	19. Ⓐ Ⓑ Ⓒ Ⓓ	34. Ⓐ Ⓑ Ⓒ Ⓓ	49. Ⓐ Ⓑ Ⓒ Ⓓ
5. Ⓐ Ⓑ Ⓒ Ⓓ	20. Ⓐ Ⓑ Ⓒ Ⓓ	35. Ⓐ Ⓑ Ⓒ Ⓓ	50. Ⓐ Ⓑ Ⓒ Ⓓ
6. Ⓐ Ⓑ Ⓒ Ⓓ	21. Ⓐ Ⓑ Ⓒ Ⓓ	36. Ⓐ Ⓑ Ⓒ Ⓓ	51. Ⓐ Ⓑ Ⓒ Ⓓ
7. Ⓐ Ⓑ Ⓒ Ⓓ	22. Ⓐ Ⓑ Ⓒ Ⓓ	37. Ⓐ Ⓑ Ⓒ Ⓓ	52. Ⓐ Ⓑ Ⓒ Ⓓ
8. Ⓐ Ⓑ Ⓒ Ⓓ	23. Ⓐ Ⓑ Ⓒ Ⓓ	38. Ⓐ Ⓑ Ⓒ Ⓓ	53. Ⓐ Ⓑ Ⓒ Ⓓ
9. Ⓐ Ⓑ Ⓒ Ⓓ	24. Ⓐ Ⓑ Ⓒ Ⓓ	39. Ⓐ Ⓑ Ⓒ Ⓓ	54. Ⓐ Ⓑ Ⓒ Ⓓ
10. Ⓐ Ⓑ Ⓒ Ⓓ	25. Ⓐ Ⓑ Ⓒ Ⓓ	40. Ⓐ Ⓑ Ⓒ Ⓓ	55. Ⓐ Ⓑ Ⓒ Ⓓ
11. Ⓐ Ⓑ Ⓒ Ⓓ	26. Ⓐ Ⓑ Ⓒ Ⓓ	41. Ⓐ Ⓑ Ⓒ Ⓓ	56. Ⓐ Ⓑ Ⓒ Ⓓ
12. Ⓐ Ⓑ Ⓒ Ⓓ	27. Ⓐ Ⓑ Ⓒ Ⓓ	42. Ⓐ Ⓑ Ⓒ Ⓓ	57. Ⓐ Ⓑ Ⓒ Ⓓ
13. Ⓐ Ⓑ Ⓒ Ⓓ	28. Ⓐ Ⓑ Ⓒ Ⓓ	43. Ⓐ Ⓑ Ⓒ Ⓓ	58. Ⓐ Ⓑ Ⓒ Ⓓ
14. Ⓐ Ⓑ Ⓒ Ⓓ	29. Ⓐ Ⓑ Ⓒ Ⓓ	44. Ⓐ Ⓑ Ⓒ Ⓓ	59. Ⓐ Ⓑ Ⓒ Ⓓ
15. Ⓐ Ⓑ Ⓒ Ⓓ	30. Ⓐ Ⓑ Ⓒ Ⓓ	45. Ⓐ Ⓑ Ⓒ Ⓓ	60. Ⓐ Ⓑ Ⓒ Ⓓ

1. Ⓐ Ⓑ Ⓒ Ⓓ   16. Ⓐ Ⓑ Ⓒ Ⓓ   31. Ⓐ Ⓑ Ⓒ Ⓓ   46. Ⓐ Ⓑ Ⓒ Ⓓ

2. Ⓐ Ⓑ Ⓒ Ⓓ   17. Ⓐ Ⓑ Ⓒ Ⓓ   32. Ⓐ Ⓑ Ⓒ Ⓓ   47. Ⓐ Ⓑ Ⓒ Ⓓ

3. Ⓐ Ⓑ Ⓒ Ⓓ   18. Ⓐ Ⓑ Ⓒ Ⓓ   33. Ⓐ Ⓑ Ⓒ Ⓓ   48. Ⓐ Ⓑ Ⓒ Ⓓ

4. Ⓐ Ⓑ Ⓒ Ⓓ   19. Ⓐ Ⓑ Ⓒ Ⓓ   34. Ⓐ Ⓑ Ⓒ Ⓓ   49. Ⓐ Ⓑ Ⓒ Ⓓ

5. Ⓐ Ⓑ Ⓒ Ⓓ   20. Ⓐ Ⓑ Ⓒ Ⓓ   35. Ⓐ Ⓑ Ⓒ Ⓓ   50. Ⓐ Ⓑ Ⓒ Ⓓ

6. Ⓐ Ⓑ Ⓒ Ⓓ   21. Ⓐ Ⓑ Ⓒ Ⓓ   36. Ⓐ Ⓑ Ⓒ Ⓓ   51. Ⓐ Ⓑ Ⓒ Ⓓ

7. Ⓐ Ⓑ Ⓒ Ⓓ   22. Ⓐ Ⓑ Ⓒ Ⓓ   37. Ⓐ Ⓑ Ⓒ Ⓓ   52. Ⓐ Ⓑ Ⓒ Ⓓ

8. Ⓐ Ⓑ Ⓒ Ⓓ   23. Ⓐ Ⓑ Ⓒ Ⓓ   38. Ⓐ Ⓑ Ⓒ Ⓓ   53. Ⓐ Ⓑ Ⓒ Ⓓ

9. Ⓐ Ⓑ Ⓒ Ⓓ   24. Ⓐ Ⓑ Ⓒ Ⓓ   39. Ⓐ Ⓑ Ⓒ Ⓓ   54. Ⓐ Ⓑ Ⓒ Ⓓ

10. Ⓐ Ⓑ Ⓒ Ⓓ   25. Ⓐ Ⓑ Ⓒ Ⓓ   40. Ⓐ Ⓑ Ⓒ Ⓓ   55. Ⓐ Ⓑ Ⓒ Ⓓ

11. Ⓐ Ⓑ Ⓒ Ⓓ   26. Ⓐ Ⓑ Ⓒ Ⓓ   41. Ⓐ Ⓑ Ⓒ Ⓓ   56. Ⓐ Ⓑ Ⓒ Ⓓ

12. Ⓐ Ⓑ Ⓒ Ⓓ   27. Ⓐ Ⓑ Ⓒ Ⓓ   42. Ⓐ Ⓑ Ⓒ Ⓓ   57. Ⓐ Ⓑ Ⓒ Ⓓ

13. Ⓐ Ⓑ Ⓒ Ⓓ   28. Ⓐ Ⓑ Ⓒ Ⓓ   43. Ⓐ Ⓑ Ⓒ Ⓓ   58. Ⓐ Ⓑ Ⓒ Ⓓ

14. Ⓐ Ⓑ Ⓒ Ⓓ   29. Ⓐ Ⓑ Ⓒ Ⓓ   44. Ⓐ Ⓑ Ⓒ Ⓓ   59. Ⓐ Ⓑ Ⓒ Ⓓ

15. Ⓐ Ⓑ Ⓒ Ⓓ   30. Ⓐ Ⓑ Ⓒ Ⓓ   45. Ⓐ Ⓑ Ⓒ Ⓓ   60. Ⓐ Ⓑ Ⓒ Ⓓ

1. Ⓐ Ⓑ Ⓒ Ⓓ 16. Ⓐ Ⓑ Ⓒ Ⓓ 31. Ⓐ Ⓑ Ⓒ Ⓓ 46. Ⓐ Ⓑ Ⓒ Ⓓ

2. Ⓐ Ⓑ Ⓒ Ⓓ 17. Ⓐ Ⓑ Ⓒ Ⓓ 32. Ⓐ Ⓑ Ⓒ Ⓓ 47. Ⓐ Ⓑ Ⓒ Ⓓ

3. Ⓐ Ⓑ Ⓒ Ⓓ 18. Ⓐ Ⓑ Ⓒ Ⓓ 33. Ⓐ Ⓑ Ⓒ Ⓓ 48. Ⓐ Ⓑ Ⓒ Ⓓ

4. Ⓐ Ⓑ Ⓒ Ⓓ 19. Ⓐ Ⓑ Ⓒ Ⓓ 34. Ⓐ Ⓑ Ⓒ Ⓓ 49. Ⓐ Ⓑ Ⓒ Ⓓ

5. Ⓐ Ⓑ Ⓒ Ⓓ 20. Ⓐ Ⓑ Ⓒ Ⓓ 35. Ⓐ Ⓑ Ⓒ Ⓓ 50. Ⓐ Ⓑ Ⓒ Ⓓ

6. Ⓐ Ⓑ Ⓒ Ⓓ 21. Ⓐ Ⓑ Ⓒ Ⓓ 36. Ⓐ Ⓑ Ⓒ Ⓓ 51. Ⓐ Ⓑ Ⓒ Ⓓ

7. Ⓐ Ⓑ Ⓒ Ⓓ 22. Ⓐ Ⓑ Ⓒ Ⓓ 37. Ⓐ Ⓑ Ⓒ Ⓓ 52. Ⓐ Ⓑ Ⓒ Ⓓ

8. Ⓐ Ⓑ Ⓒ Ⓓ 23. Ⓐ Ⓑ Ⓒ Ⓓ 38. Ⓐ Ⓑ Ⓒ Ⓓ 53. Ⓐ Ⓑ Ⓒ Ⓓ

9. Ⓐ Ⓑ Ⓒ Ⓓ 24. Ⓐ Ⓑ Ⓒ Ⓓ 39. Ⓐ Ⓑ Ⓒ Ⓓ 54. Ⓐ Ⓑ Ⓒ Ⓓ

10. Ⓐ Ⓑ Ⓒ Ⓓ 25. Ⓐ Ⓑ Ⓒ Ⓓ 40. Ⓐ Ⓑ Ⓒ Ⓓ 55. Ⓐ Ⓑ Ⓒ Ⓓ

11. Ⓐ Ⓑ Ⓒ Ⓓ 26. Ⓐ Ⓑ Ⓒ Ⓓ 41. Ⓐ Ⓑ Ⓒ Ⓓ 56. Ⓐ Ⓑ Ⓒ Ⓓ

12. Ⓐ Ⓑ Ⓒ Ⓓ 27. Ⓐ Ⓑ Ⓒ Ⓓ 42. Ⓐ Ⓑ Ⓒ Ⓓ 57. Ⓐ Ⓑ Ⓒ Ⓓ

13. Ⓐ Ⓑ Ⓒ Ⓓ 28. Ⓐ Ⓑ Ⓒ Ⓓ 43. Ⓐ Ⓑ Ⓒ Ⓓ 58. Ⓐ Ⓑ Ⓒ Ⓓ

14. Ⓐ Ⓑ Ⓒ Ⓓ 29. Ⓐ Ⓑ Ⓒ Ⓓ 44. Ⓐ Ⓑ Ⓒ Ⓓ 59. Ⓐ Ⓑ Ⓒ Ⓓ

15. Ⓐ Ⓑ Ⓒ Ⓓ 30. Ⓐ Ⓑ Ⓒ Ⓓ 45. Ⓐ Ⓑ Ⓒ Ⓓ 60. Ⓐ Ⓑ Ⓒ Ⓓ

1. Ⓐ Ⓑ Ⓒ Ⓓ   16. Ⓐ Ⓑ Ⓒ Ⓓ   31. Ⓐ Ⓑ Ⓒ Ⓓ   46. Ⓐ Ⓑ Ⓒ Ⓓ

2. Ⓐ Ⓑ Ⓒ Ⓓ   17. Ⓐ Ⓑ Ⓒ Ⓓ   32. Ⓐ Ⓑ Ⓒ Ⓓ   47. Ⓐ Ⓑ Ⓒ Ⓓ

3. Ⓐ Ⓑ Ⓒ Ⓓ   18. Ⓐ Ⓑ Ⓒ Ⓓ   33. Ⓐ Ⓑ Ⓒ Ⓓ   48. Ⓐ Ⓑ Ⓒ Ⓓ

4. Ⓐ Ⓑ Ⓒ Ⓓ   19. Ⓐ Ⓑ Ⓒ Ⓓ   34. Ⓐ Ⓑ Ⓒ Ⓓ   49. Ⓐ Ⓑ Ⓒ Ⓓ

5. Ⓐ Ⓑ Ⓒ Ⓓ   20. Ⓐ Ⓑ Ⓒ Ⓓ   35. Ⓐ Ⓑ Ⓒ Ⓓ   50. Ⓐ Ⓑ Ⓒ Ⓓ

6. Ⓐ Ⓑ Ⓒ Ⓓ   21. Ⓐ Ⓑ Ⓒ Ⓓ   36. Ⓐ Ⓑ Ⓒ Ⓓ   51. Ⓐ Ⓑ Ⓒ Ⓓ

7. Ⓐ Ⓑ Ⓒ Ⓓ   22. Ⓐ Ⓑ Ⓒ Ⓓ   37. Ⓐ Ⓑ Ⓒ Ⓓ   52. Ⓐ Ⓑ Ⓒ Ⓓ

8. Ⓐ Ⓑ Ⓒ Ⓓ   23. Ⓐ Ⓑ Ⓒ Ⓓ   38. Ⓐ Ⓑ Ⓒ Ⓓ   53. Ⓐ Ⓑ Ⓒ Ⓓ

9. Ⓐ Ⓑ Ⓒ Ⓓ   24. Ⓐ Ⓑ Ⓒ Ⓓ   39. Ⓐ Ⓑ Ⓒ Ⓓ   54. Ⓐ Ⓑ Ⓒ Ⓓ

10. Ⓐ Ⓑ Ⓒ Ⓓ   25. Ⓐ Ⓑ Ⓒ Ⓓ   40. Ⓐ Ⓑ Ⓒ Ⓓ   55. Ⓐ Ⓑ Ⓒ Ⓓ

11. Ⓐ Ⓑ Ⓒ Ⓓ   26. Ⓐ Ⓑ Ⓒ Ⓓ   41. Ⓐ Ⓑ Ⓒ Ⓓ   56. Ⓐ Ⓑ Ⓒ Ⓓ

12. Ⓐ Ⓑ Ⓒ Ⓓ   27. Ⓐ Ⓑ Ⓒ Ⓓ   42. Ⓐ Ⓑ Ⓒ Ⓓ   57. Ⓐ Ⓑ Ⓒ Ⓓ

13. Ⓐ Ⓑ Ⓒ Ⓓ   28. Ⓐ Ⓑ Ⓒ Ⓓ   43. Ⓐ Ⓑ Ⓒ Ⓓ   58. Ⓐ Ⓑ Ⓒ Ⓓ

14. Ⓐ Ⓑ Ⓒ Ⓓ   29. Ⓐ Ⓑ Ⓒ Ⓓ   44. Ⓐ Ⓑ Ⓒ Ⓓ   59. Ⓐ Ⓑ Ⓒ Ⓓ

15. Ⓐ Ⓑ Ⓒ Ⓓ   30. Ⓐ Ⓑ Ⓒ Ⓓ   45. Ⓐ Ⓑ Ⓒ Ⓓ   60. Ⓐ Ⓑ Ⓒ Ⓓ

1. Ⓐ Ⓑ Ⓒ Ⓓ	16. Ⓐ Ⓑ Ⓒ Ⓓ	31. Ⓐ Ⓑ Ⓒ Ⓓ	46. Ⓐ Ⓑ Ⓒ Ⓓ
2. Ⓐ Ⓑ Ⓒ Ⓓ	17. Ⓐ Ⓑ Ⓒ Ⓓ	32. Ⓐ Ⓑ Ⓒ Ⓓ	47. Ⓐ Ⓑ Ⓒ Ⓓ
3. Ⓐ Ⓑ Ⓒ Ⓓ	18. Ⓐ Ⓑ Ⓒ Ⓓ	33. Ⓐ Ⓑ Ⓒ Ⓓ	48. Ⓐ Ⓑ Ⓒ Ⓓ
4. Ⓐ Ⓑ Ⓒ Ⓓ	19. Ⓐ Ⓑ Ⓒ Ⓓ	34. Ⓐ Ⓑ Ⓒ Ⓓ	49. Ⓐ Ⓑ Ⓒ Ⓓ
5. Ⓐ Ⓑ Ⓒ Ⓓ	20. Ⓐ Ⓑ Ⓒ Ⓓ	35. Ⓐ Ⓑ Ⓒ Ⓓ	50. Ⓐ Ⓑ Ⓒ Ⓓ
6. Ⓐ Ⓑ Ⓒ Ⓓ	21. Ⓐ Ⓑ Ⓒ Ⓓ	36. Ⓐ Ⓑ Ⓒ Ⓓ	51. Ⓐ Ⓑ Ⓒ Ⓓ
7. Ⓐ Ⓑ Ⓒ Ⓓ	22. Ⓐ Ⓑ Ⓒ Ⓓ	37. Ⓐ Ⓑ Ⓒ Ⓓ	52. Ⓐ Ⓑ Ⓒ Ⓓ
8. Ⓐ Ⓑ Ⓒ Ⓓ	23. Ⓐ Ⓑ Ⓒ Ⓓ	38. Ⓐ Ⓑ Ⓒ Ⓓ	53. Ⓐ Ⓑ Ⓒ Ⓓ
9. Ⓐ Ⓑ Ⓒ Ⓓ	24. Ⓐ Ⓑ Ⓒ Ⓓ	39. Ⓐ Ⓑ Ⓒ Ⓓ	54. Ⓐ Ⓑ Ⓒ Ⓓ
10. Ⓐ Ⓑ Ⓒ Ⓓ	25. Ⓐ Ⓑ Ⓒ Ⓓ	40. Ⓐ Ⓑ Ⓒ Ⓓ	55. Ⓐ Ⓑ Ⓒ Ⓓ
11. Ⓐ Ⓑ Ⓒ Ⓓ	26. Ⓐ Ⓑ Ⓒ Ⓓ	41. Ⓐ Ⓑ Ⓒ Ⓓ	56. Ⓐ Ⓑ Ⓒ Ⓓ
12. Ⓐ Ⓑ Ⓒ Ⓓ	27. Ⓐ Ⓑ Ⓒ Ⓓ	42. Ⓐ Ⓑ Ⓒ Ⓓ	57. Ⓐ Ⓑ Ⓒ Ⓓ
13. Ⓐ Ⓑ Ⓒ Ⓓ	28. Ⓐ Ⓑ Ⓒ Ⓓ	43. Ⓐ Ⓑ Ⓒ Ⓓ	58. Ⓐ Ⓑ Ⓒ Ⓓ
14. Ⓐ Ⓑ Ⓒ Ⓓ	29. Ⓐ Ⓑ Ⓒ Ⓓ	44. Ⓐ Ⓑ Ⓒ Ⓓ	59. Ⓐ Ⓑ Ⓒ Ⓓ
15. Ⓐ Ⓑ Ⓒ Ⓓ	30. Ⓐ Ⓑ Ⓒ Ⓓ	45. Ⓐ Ⓑ Ⓒ Ⓓ	60. Ⓐ Ⓑ Ⓒ Ⓓ

1. Ⓐ Ⓑ Ⓒ Ⓓ	16. Ⓐ Ⓑ Ⓒ Ⓓ	31. Ⓐ Ⓑ Ⓒ Ⓓ	46. Ⓐ Ⓑ Ⓒ Ⓓ
2. Ⓐ Ⓑ Ⓒ Ⓓ	17. Ⓐ Ⓑ Ⓒ Ⓓ	32. Ⓐ Ⓑ Ⓒ Ⓓ	47. Ⓐ Ⓑ Ⓒ Ⓓ
3. Ⓐ Ⓑ Ⓒ Ⓓ	18. Ⓐ Ⓑ Ⓒ Ⓓ	33. Ⓐ Ⓑ Ⓒ Ⓓ	48. Ⓐ Ⓑ Ⓒ Ⓓ
4. Ⓐ Ⓑ Ⓒ Ⓓ	19. Ⓐ Ⓑ Ⓒ Ⓓ	34. Ⓐ Ⓑ Ⓒ Ⓓ	49. Ⓐ Ⓑ Ⓒ Ⓓ
5. Ⓐ Ⓑ Ⓒ Ⓓ	20. Ⓐ Ⓑ Ⓒ Ⓓ	35. Ⓐ Ⓑ Ⓒ Ⓓ	50. Ⓐ Ⓑ Ⓒ Ⓓ
6. Ⓐ Ⓑ Ⓒ Ⓓ	21. Ⓐ Ⓑ Ⓒ Ⓓ	36. Ⓐ Ⓑ Ⓒ Ⓓ	51. Ⓐ Ⓑ Ⓒ Ⓓ
7. Ⓐ Ⓑ Ⓒ Ⓓ	22. Ⓐ Ⓑ Ⓒ Ⓓ	37. Ⓐ Ⓑ Ⓒ Ⓓ	52. Ⓐ Ⓑ Ⓒ Ⓓ
8. Ⓐ Ⓑ Ⓒ Ⓓ	23. Ⓐ Ⓑ Ⓒ Ⓓ	38. Ⓐ Ⓑ Ⓒ Ⓓ	53. Ⓐ Ⓑ Ⓒ Ⓓ
9. Ⓐ Ⓑ Ⓒ Ⓓ	24. Ⓐ Ⓑ Ⓒ Ⓓ	39. Ⓐ Ⓑ Ⓒ Ⓓ	54. Ⓐ Ⓑ Ⓒ Ⓓ
10. Ⓐ Ⓑ Ⓒ Ⓓ	25. Ⓐ Ⓑ Ⓒ Ⓓ	40. Ⓐ Ⓑ Ⓒ Ⓓ	55. Ⓐ Ⓑ Ⓒ Ⓓ
11. Ⓐ Ⓑ Ⓒ Ⓓ	26. Ⓐ Ⓑ Ⓒ Ⓓ	41. Ⓐ Ⓑ Ⓒ Ⓓ	56. Ⓐ Ⓑ Ⓒ Ⓓ
12. Ⓐ Ⓑ Ⓒ Ⓓ	27. Ⓐ Ⓑ Ⓒ Ⓓ	42. Ⓐ Ⓑ Ⓒ Ⓓ	57. Ⓐ Ⓑ Ⓒ Ⓓ
13. Ⓐ Ⓑ Ⓒ Ⓓ	28. Ⓐ Ⓑ Ⓒ Ⓓ	43. Ⓐ Ⓑ Ⓒ Ⓓ	58. Ⓐ Ⓑ Ⓒ Ⓓ
14. Ⓐ Ⓑ Ⓒ Ⓓ	29. Ⓐ Ⓑ Ⓒ Ⓓ	44. Ⓐ Ⓑ Ⓒ Ⓓ	59. Ⓐ Ⓑ Ⓒ Ⓓ
15. Ⓐ Ⓑ Ⓒ Ⓓ	30. Ⓐ Ⓑ Ⓒ Ⓓ	45. Ⓐ Ⓑ Ⓒ Ⓓ	60. Ⓐ Ⓑ Ⓒ Ⓓ

1. Ⓐ Ⓑ Ⓒ Ⓓ 16. Ⓐ Ⓑ Ⓒ Ⓓ 31. Ⓐ Ⓑ Ⓒ Ⓓ 46. Ⓐ Ⓑ Ⓒ Ⓓ

2. Ⓐ Ⓑ Ⓒ Ⓓ 17. Ⓐ Ⓑ Ⓒ Ⓓ 32. Ⓐ Ⓑ Ⓒ Ⓓ 47. Ⓐ Ⓑ Ⓒ Ⓓ

3. Ⓐ Ⓑ Ⓒ Ⓓ 18. Ⓐ Ⓑ Ⓒ Ⓓ 33. Ⓐ Ⓑ Ⓒ Ⓓ 48. Ⓐ Ⓑ Ⓒ Ⓓ

4. Ⓐ Ⓑ Ⓒ Ⓓ 19. Ⓐ Ⓑ Ⓒ Ⓓ 34. Ⓐ Ⓑ Ⓒ Ⓓ 49. Ⓐ Ⓑ Ⓒ Ⓓ

5. Ⓐ Ⓑ Ⓒ Ⓓ 20. Ⓐ Ⓑ Ⓒ Ⓓ 35. Ⓐ Ⓑ Ⓒ Ⓓ 50. Ⓐ Ⓑ Ⓒ Ⓓ

6. Ⓐ Ⓑ Ⓒ Ⓓ 21. Ⓐ Ⓑ Ⓒ Ⓓ 36. Ⓐ Ⓑ Ⓒ Ⓓ 51. Ⓐ Ⓑ Ⓒ Ⓓ

7. Ⓐ Ⓑ Ⓒ Ⓓ 22. Ⓐ Ⓑ Ⓒ Ⓓ 37. Ⓐ Ⓑ Ⓒ Ⓓ 52. Ⓐ Ⓑ Ⓒ Ⓓ

8. Ⓐ Ⓑ Ⓒ Ⓓ 23. Ⓐ Ⓑ Ⓒ Ⓓ 38. Ⓐ Ⓑ Ⓒ Ⓓ 53. Ⓐ Ⓑ Ⓒ Ⓓ

9. Ⓐ Ⓑ Ⓒ Ⓓ 24. Ⓐ Ⓑ Ⓒ Ⓓ 39. Ⓐ Ⓑ Ⓒ Ⓓ 54. Ⓐ Ⓑ Ⓒ Ⓓ

10. Ⓐ Ⓑ Ⓒ Ⓓ 25. Ⓐ Ⓑ Ⓒ Ⓓ 40. Ⓐ Ⓑ Ⓒ Ⓓ 55. Ⓐ Ⓑ Ⓒ Ⓓ

11. Ⓐ Ⓑ Ⓒ Ⓓ 26. Ⓐ Ⓑ Ⓒ Ⓓ 41. Ⓐ Ⓑ Ⓒ Ⓓ 56. Ⓐ Ⓑ Ⓒ Ⓓ

12. Ⓐ Ⓑ Ⓒ Ⓓ 27. Ⓐ Ⓑ Ⓒ Ⓓ 42. Ⓐ Ⓑ Ⓒ Ⓓ 57. Ⓐ Ⓑ Ⓒ Ⓓ

13. Ⓐ Ⓑ Ⓒ Ⓓ 28. Ⓐ Ⓑ Ⓒ Ⓓ 43. Ⓐ Ⓑ Ⓒ Ⓓ 58. Ⓐ Ⓑ Ⓒ Ⓓ

14. Ⓐ Ⓑ Ⓒ Ⓓ 29. Ⓐ Ⓑ Ⓒ Ⓓ 44. Ⓐ Ⓑ Ⓒ Ⓓ 59. Ⓐ Ⓑ Ⓒ Ⓓ

15. Ⓐ Ⓑ Ⓒ Ⓓ 30. Ⓐ Ⓑ Ⓒ Ⓓ 45. Ⓐ Ⓑ Ⓒ Ⓓ 60. Ⓐ Ⓑ Ⓒ Ⓓ